How to Foster Dogs

Fr

H

Pat M

Dogwise™ Publishing

Wenatchee, Washington U.S.A.

How to Foster Dogs
From Homeless to Homeward Bound
Pat Miller, CPDT-KA, CBCC-KA

Dogwise Publishing
A Division of Direct Book Service, Inc.
403 South Mission Street, Wenatchee, Washington 98801
1-509-663-9115, 1-800-776-2665
www.dogwisepublishing.com / info@dogwisepublishing.com

Photos: Greg Armes, Sumner Fowler, Shirley Greenlief, Claudia Husemann, Nancy Kerns, Pat Miller, Paul Miller, Shannon McAuliffe
Graphic design: Lindsay Peternell
Cover design: Kristy Allen

Limits of Liability and Disclaimer of Warranty:
The author and publisher shall not be liable in the event of incidental or consequential damages in connection with, or arising out of, the furnishing, performance, or use of the instructions and suggestions contained in this book.

Library of Congress Cataloging-in-Publication Data
Miller, Pat, 1951 October 14-
  How to foster dogs : from homeless to homeward bound / by Pat Miller, CPDT-KA, CBCC-KA.
     pages cm
  Includes index.
  ISBN 978-1-61781-134-0
  1. Dogs. 2. Foster care of animals. 3. Dogs--Behavior. I. Title.
  SF427.M587 2013
  636.7'0832--dc23
                                              2013039775

ISBN: 978-1-61781-134-0

Printed in the U.S.A.

More praise for *How to Foster Dogs*

More people than ever are interested in fostering dogs, however this romantic notion may be a disaster without the right tools, which is what a Certified Dog Behavior Consultant provides. Miller's tips will help you to integrate the Foster pup into your family from day one, and without that often talked about need to dominate the dog—as she effectively busts the so-called dominance theory. Miller also addresses what to do regarding training and several common behavioral issues.

Steve Dale, CABC, radio host (Steve Dale's *Pet World,* syndicated Black Dog Radio Productions and WGN Radio); newspaper column *My Pet World* (Tribune Content Agency); contributing editor USA Weekend; contributor to *Decoding Your Dog*

It is a realistic but enthusiastic and informative look at fostering dogs, a truly lifesaving and rewarding strategy to help shelters reduce euthanasia of healthy and treatable pets. The materials are comprehensive with helpful citations for the behavior/training information and health information. I especially appreciated the discussion of all the different types of animal welfare groups that use fostering and Pat is smart to recommend doing a little research on any group foster volunteers' endeavor to help—you just never know until you see it firsthand.

Sharon M. Harmon, CAWA, Executive Director Oregon Humane Society

Pat Miller does an excellent, comprehensive job of teaching readers how to go about fostering intelligently and successfully. Invaluable information and support is plentiful on every step of the journey, beginning with finding a foster dog, making preparations and conducting introductions, through training and socialization and, finally, the adoption process. *How to Foster Dogs* is a vital resource that should be required reading for every foster home!

Nicole Wilde, author *Help for Your Fearful Dog* and *Don't Leave Me! Step by Step Help for Your Dog's Separation Anxiety*

This clearly written, comprehensive guide covers everything from how to find a shelter or rescue to partner with, to how to decide which dogs you can take on, to the heartachy joy of saying goodbye at adoption. Of course, all that is just what I'd expect given Pat's decades of experience in sheltering and rescue.

Jolanta Benal, CPDT-KA, CBCC-KA, author *The Dog Trainer's Complete Guide to a Happy, Well-Behaved Pet* and host of *The Dog Trainer's Quick and Dirty Tips for Teaching and Caring for Your Pet*

Dedication

To Mandy, Squid, Captain Jack Cricket, Boing, and all the others who over the years have shared our home and our hearts on their way to their forever homes.

The lovely and perfect Mandy, with me, many long years ago. Photo: Sumner Fowler

Table of Contents

Acknowledgements

Writing a book is never a single-person endeavor. My thanks go to the many Peaceable Paws apprentices and academy graduates who offered suggestions for this book including Catherine Schuler, Helene Goldberger, Jolanta Benal, Sean Howard, Simone deLima, Laura Nalven, Melissa Chow, Bob Ryder, Valerie Balwanz, Amy Trice, Andrea Brady, Sharon Messersmith, Diana Foley, Estelina Dallett, Leslie Clifton, Chris Danker and AnnMarie Easton. My apologies to any I may have inadvertently left out.

And of course, eternal thanks to my husband, Paul Miller, for his never-ending support for this book as well as all my countless other projects and a life of total immersion in animals. I could not do this without him.

Finally, thanks, always, to the people at Dogwise Publishing for being such a fantastic resource for so many of the excellent new books that serve to feed our ravenous appetite for new information about our beloved dogs. We are incredibly lucky to have these great folks.

Introduction

In the 1980s and 1990s, when I was working at the Marin Humane Society (MHS) in Novato, California, fostering was a very occasional thing. Every once in a while an animal in need would touch the heart of a shelter worker who offered to take that dog or cat home to help her through a bout of upper respiratory infection, or some other temporary medical issue. Occasionally someone would foster a pregnant mom until she could have her babies, or foster a litter of puppies or kittens until they were old enough to be placed, but in those days, with euthanasia rates in the U.S. at an all time high (18 to 20 million homeless dogs and cats euthanized at shelters every year), it seemed to make little sense to save unborn puppies, or to allocate many of our limited resources to save dogs who had medical problems. Dogs with behavioral problems were rarely fostered. Resources were always scarce, and there were inevitably always more healthy dogs in the shelter's kennels to take the place of sick ones, or those who were behaviorally challenged.

I fostered several dogs and a few cats over the 20 years I was at MHS. Most memorable for me was Mandy, a seven-year-old spayed female tri-color Collie, who was surrendered by her owner because she was incontinent. In addition, she was seriously overweight, badly matted, and had infected burns on the insides of her hind legs where the urine had soaked through her mats to the skin.

*A piece of my heart will forever rest at the
Marin Humane Society. Photo: Paul Miller*

I had Rough Collies (Lassie Collies) throughout my childhood, and am still drawn to them to this day. My heart went out to Mandy, and I offered to foster her. Our shelter vet treated the urine burns and provided medication for her incontinence. I took her home; she walked into my house and lay down on the floor like she had lived there all her life. She never left my family. A "foster failure," as some call it when a foster parent ends up adopting a foster charge. I prefer to call it a foster success.

We have fostered numerous other shelter animals since those days. It was one of the things that kept me sane and allowed me to work at a shelter for two decades—knowing that, from time to time, I was able to make a difference to at least one of the many.

These days fostering is no longer an occasional intervention—it's a booming industry. With national euthanasia numbers down to a relatively low (but still too high) three to four million dogs and cats per year, it makes sense to work harder to save as many as possible. While the admirable goal of a real no-kill society is still some distance off, there is a glimmer of light at the end of the tunnel. Every dog who goes to a responsible foster home from a responsible shelter or rescue group makes that light shine a little bit brighter.

My hope is that this book will help to brighten that light even more by providing foster parents, shelters and rescues with a resource to create more foster successes, whether adopted by the fosterer or placed in some other lifelong loving home. Along with exceptional fostering efforts, this will require a societal mind shift toward greater regulation of breeding and retail sales of dogs—especially from puppy mills and Internet brokers—so the supply no longer outweighs the demand. I would love for all of us to see, in our lifetime, a world where every dog and cat born is valued, loved and cared for to old age. It's the least we can do for the animal companions who share our homes and our hearts.

1

A Fostering Overview

Squid was the cutest seven-week-old puppy you could ever imagine—probably a Jack Russell Terrier mix, white and tan with a rakish brown spot over his left ear and another covering his left eye. He was also on the euthanasia list at our local animal shelter. He had failed his assessment for aggression. Shelter staff asked me if I would work with him under the shelter's Gold Paw behavioral foster care partnership program. After spending a little time with him, I agreed. I felt his frustration-aggression and lack of impulse control would be easily modifiable behaviors, especially given his young age, adorable appearance, and the fact that that he had the potential to be highly adoptable. Time would tell.

Foster pup Squid, slated for euthanasia at our local shelter until he was accepted into the Gold Paw foster program.

What, exactly, is fostering?

An online dictionary defines fostering as:

1. to promote the growth or development of; further; encourage

2. to bring up, raise, or rear, as a foster child

3. to care for or cherish.

All three of those definitions apply to the world of canine fostering. More specifically, with animal fostering we are usually referring to an animal from a shelter or rescue group—perhaps originally a stray, perhaps surrendered by an owner, or perhaps a victim from an animal neglect or cruelty case. Shelters often seek out foster homes for puppies who are too young to thrive in a shelter environment, dogs who have medical or behavioral needs that are better met outside a shelter, or simply because they have too many dogs for the available kennel space in the shelter. Even pregnant females may need fostering!

In any case, the purpose of fostering is to look for alternatives to euthanasia, increase the live-release rate, and thereby reduce the number of animals who die at shelters due to homelessness. Many rescue groups are unsheltered (meaning they have no one place to house dogs), and rely primarily on foster homes to do their good work. Sometimes these groups end up paying to board dogs at kennels when foster homes are in short supply.

Fostering implies a limited time commitment, a temporary arrangement in which a person agrees to house and take care of the dog until a permanent home can be found. Fostering is not the same as adopting, where a person agrees to take ownership of the dog on a permanent basis. Of course, sometimes the person agreeing to foster the dog ends up adopting the dog!

Yet another foster opportunity exists for dogs belonging to military personnel who must go overseas and cannot take their beloved canine companions with them. In the past many of those dogs were sadly rehomed. Today, a growing number of our troops are comforted knowing that responsible fosterers are caring for their dogs, and they will be greeting with tail wags and loving wet kisses when they return home.

In this book we will focus on what can be termed a "formal" fostering arrangement between a shelter or a rescue group and the person who agrees to foster the dog. However, "informal" fostering arrangements happen all the time as well. A relative might ask you to care for a dog for a couple of months while she travels abroad. You might pick up a stray dog and house him for a couple weeks while attempting to locate the owner. In these latter cases there is no formal arrangement, but what you do in terms of caring for the dog and introducing him to your own dogs and/or human family is pretty much the same.

Unlike "informal" fostering situations, in many cases there will be some sort of written agreement between you and the shelter or rescue group detailing the responsibilities of both parties. You will want to know who you are dealing with and what you can expect from them in terms of guidance and financial support. And, of course, the organization will want to be convinced that you have the skills and desire to provide a good fostering home for the dog in question. We will cover these arrangements in more detail below and in the next chapter.

Why foster?

You foster because you care. You see it as a way you can make a difference in a world that sometimes doesn't care enough. Perhaps you want to pay tribute to a past animal companion who is no longer with you. You may realize your lifestyle doesn't allow for longtime commitments to a furry family member, but you are able to make a shorter-term commitment to a dog who might not otherwise have a chance. In the end, you do it because you care.

Perhaps the more pertinent question is: *should* you foster? A lot of factors play into a well-considered decision to become a canine foster parent. Here are some things you will want to give serious thought to prior to taking on the responsibility of fostering one or more dogs:

1. **Disruption to your domestic tranquility.** Some fosters, like Mandy, the Collie I describe in the Introduction, are no bother at all. They blend into the woodwork like they've always been there. More frequently, however, foster dogs are likely to come with lots of energy and a potential for behavior challenges. Remember that homeless dogs were given up for a reason—and the number one reason dogs are given up or not reclaimed is behavior. Your foster's behavior may have been something her prior owner couldn't live with. If you have dogs or other companion animals of your own, the presence of non-resident dogs can affect *their* quality of life, as well as yours. Make sure you are ready for the impact this may have on your lifestyle and serenity. (By the way—if your own dogs don't do well with new dogs in their home, don't even *think* about fostering dogs. If you have your heart set on fostering, consider other species that your dogs will tolerate well. If you have cats, use extreme caution when bringing a foster dog into your home, until you know your cats will be safe.)

2. **Environmental factors.** Where do you plan to keep your foster dog? Will he be crated at night? (Is he crate trained? Will he tolerate a crate? Can you crate train him?) Will he share your bedroom with your dogs and spouse? Your bed?

(Generally not recommended for a foster, since his future forever home may not want him in the bed, and it might be a hard habit to change.) Where will he be during the day when no one is home? Can you take him to work with you? What if he is destructive? What if he barks a lot?

3. **Family buy-in.** It's important that your entire family is on board with the fostering project. Anger or resentment over a canine intruder can fester and damage human relationships, and may result in actual abuse to the dog, if there are family members who are unhappy about fostering. You cannot assume they will get over it—the entire family needs to be positive about fostering *before* you bring a dog home.

4. **Financial considerations.** Some shelters and rescues will pay all expenses for your foster dog. Others will pay some, while some expect their foster parents to bear the entire financial burden of fostering. A tiny minority actually pay their foster homes a small weekly stipend. Make sure you are clear about finances before agreeing to foster—and make sure your own finances can weather the cost, if that's the arrangement.

5. **Know your limits.** We still live in a world where there are *many* more dogs than there are homes. You could foster every single dog in your local animal shelter today, and there would be more tomorrow. And the next day. And the next. An increasing number of horrendous hoarder cases are reported in the news weekly, involving well-intentioned rescue groups and foster homes. Don't let yourself become one of these. Know your limits, and have one or more people you love and trust lined up to let you know if they think you are going beyond reasonable limits. You really aren't helping any dogs if you take on more than you can provide for.

The supply of homeless dogs is endless.
You can't take them all…know your limits!

6. **Legal considerations.** In today's litigious world, legal considerations must be taken seriously. Does your chosen group's insurance cover you as a volunteer if your foster dog bites someone, or causes some other accident or injury? (Hint—they should.) If not, does your homeowner's insurance cover you? Are you being asked to foster a breed of dog that your insurance might exclude from coverage? Better to know the answers to these questions *before* there's an incident, rather than find out afterward that you aren't covered.

7. **Record keeping.** You may be required to keep records of your foster dog's medical and behavioral history during the time he spends with you. (If you aren't required to do so, you should, anyway. It will greatly facilitate the dog's transition to his new home. More on this in Chapter 3.)

8. **Foster failures.** True foster failures are those where the dog doesn't adapt well to his foster home, where the foster family doesn't have the skills or patience to work with his behavior issues, or worse, where your foster ends up not being a good candidate for adoption. If it's simply that your home is an unsuitable environment for that particular dog, then hopefully there's another foster home available that is better suited for him, and another foster dog who is better suited for yours. If it becomes evident that the dog is not a good adoption candidate, either for medical or behavioral reasons, you will need to deal with the strong emotions that are inevitable—and normal—if euthanasia becomes the appropriate outcome for your foster. Even lifetime animal protection professionals struggle with the emotional issues of euthanasia. It's not easy, nor should it be. You should only consider fostering if you are prepared to face this hopefully rare possibility. (More on this in Chapter 3.)

Who needs foster volunteers?

Almost every 501(c)3 non-profit animal shelter and rescue group now makes extensive use of foster homes, and many municipal (government-run) animal care and control programs do as well. You should easily be able to find a group in your area who needs the services of a good foster home. An online search for "dog, foster home, volunteer" will bring up a long list of such organizations anywhere in the country. You can also look in the yellow pages of your local telephone book (or online Yellow Pages) to find shelter and rescue groups who are looking for foster homes. It's important to remember that your fostering experience needs to be a mutually satisfying one. Just like a job interview, you want to be sure your future foster "employer" will be a good fit in the areas of fostering philosophies, training and behavior modification tools and methods, emotional and financial support, and basic animal care.

2

Ready, Set, Go!

Finding a foster program

Your emotional fortitude and your own philosophical stance on rescuing will guide you in deciding what agency, or agencies, you would like to partner with. A group's mission, charter, policies and procedures will help you decide whether you are comfortable working with them or not.

There are shelters and rescue groups that accept virtually all animals brought to them. These groups, called "full service" or "open admission" agencies, will of necessity (assuming they are responsible caretakers) euthanize a greater percentage of the animals they take in than groups that are selective about the dogs they accept. Because they don't discriminate on intake, open admission agencies inevitably find themselves caring for dogs who are too injured, too ill or too behaviorally damaged to become healthy, safe or reasonable companions. Limited admission or selective intake groups (that sometimes use the unfortunate "no-kill" label), because they can choose which dogs they accept, euthanize a much smaller percentage of the dogs they care for. Most, however—if they are responsible caretakers and honest about it—will euthanize at some point if a dog in their care deteriorates behaviorally or physically, such that it is no longer humane to keep the dog, or it becomes apparent that he isn't safe to place in the community.

The sheltering industry has accepted a definition of "no-kill" that most of the general public doesn't understand—that *adoptable* dogs aren't euthanized. But which illnesses, injuries or behavior challenges a shelter has the resources, willingness and ability to treat varies

widely from one to the next. So a dog who is *unadoptable* and euthanized for upper respiratory infection at one so-called no-kill shelter
might well have been declared *adoptable*, treated and rehomed at the
next. A dog with a broken leg might be treated at one, and not at the
other. A dog with resource guarding behavior might be euthanized at
one shelter, rehomed at the next, and rehabilitated and rehomed at a
third. They can all call themselves "no-kill." (Hence my dislike of the
use of this misleading term.)

*A dog shouldn't have to live in a small shelter kennel for
years on end. There are fates worse than a gentle death.*

Another caveat of low- or no-kill shelters is that *some* facilities keep
dogs in their kennels far longer than is appropriate or humane—
many months, even many years. It's well-accepted in the behavior
field that a kennel environment can be highly stressful for some dogs,
and that the resulting stress can cause a multitude of serious behavior problems, including aggression and compulsive behaviors such as

spinning and self-mutilation. These behaviors can be very difficult to resolve, and may end up causing heartbreak for a foster parent or eventual adopter.

Additionally, dogs impounded in neglect and cruelty (including hoarder) cases are often very behaviorally damaged. In the past, most seriously behaviorally damaged dogs were euthanized by the impounding shelters. Now, with the growth of the "no-kill" movement and the rise of the fostering industry, more of these dogs are being rescued—and fostered. The good news is that we are finding some of the dogs once believed damaged beyond repair actually can recover, with adequate and appropriate behavior modification. The bad news is that while many of them still cannot be repaired, many of the very badly damaged dogs are still being fostered and eventually placed in homes with warmhearted loving humans who have no idea what they are getting themselves into. They often think that sheer love will be enough to help the dog become "normal." It won't. The result is heartbreak for foster parents as well as for the adopters.

A recent online discussion of no-kill shelters and rescues elicited many comments including the following from several dog lovers who are involved in rescuing and fostering:

- "It's an emotional issue but at the end of the day, you have to weigh the good of the whole rescue group and the number of animals you are saving against one animal that you cannot save."

- "Letting an animal live just so you can say you are no-kill but keeping him in an environment that does not provide a quality of life is cruel to that animal."

- "I think the 'no-kill' movement has gone way too far the wrong way. I am seeing dogs in pet homes that are scary time bombs just waiting to go off, and owners who have no idea. I have one in training at the moment. It is so damaged that one day it will kill another dog. I am sure of it, but the owner is in total denial."

- "Yes, the 'no-kill' movement will backfire. Dogs are adopted out to unsuspecting, naive people who just want a nice companion. Society will become ever more intolerant of dogs as more bites and fights occur. More landlords will say no, more lawyers will be employed. More municipalities and counties will outlaw dogs. People will choose not to adopt from shelters or rescues, based on bad experiences or word of mouth."

Saving lives is admirable. Keeping dogs alive who have no hope of a decent life, or who are very likely to cause injury or heartbreak, is not.

A shelter/rescue primer

Many people think that shelters must all come under some national governing body that regulates what they do, a universal "mother club" like Red Cross, Boy Scouts and Girl Scouts. In fact, the exact opposite is true—with a few rare exceptions, every shelter is its own entity, complete within itself, with its own policies and procedures, its own governing body, and its own list of services offered—or not offered.

National groups like the Humane Society of the United States (HSUS), American Humane Association, United Animal Nations and the American Society for the Prevention of Cruelty to Animals (ASPCA) do little to dispel the confusion. None of these organizations has anything to do with the management of shelters around the country—they are primarily *educational* organizations, offering training, materials and educational conferences for a fee to local shelters, and issue-based information to the public. (The ASPCA does have a shelter in New York City, and HSUS has a large-animal rescue facility.) Some have offices around the United States, or sometimes the world; many of them participate in disaster response efforts; and some are heavily involved in lawmaking, sometimes pursuing legislation whether local agencies support it or not. Well-meaning animal lovers often join and support these national organizations, believing that donation dollars sent to those groups somehow find their way back to help animals in shelters in their own communities. This is rarely the case—once in a great while, during a disaster or a high-profile cruelty case perhaps. In most cases, however, rarely a penny goes to assisting with the day-to-day costs of feeding and caring for sheltered animals.

Types of shelters

Although every shelter is its own entity, you can group them into similar types according to their structure and the services they offer:

Municipal shelter. This type of shelter is owned and run by your local government—city or county, with names like "San Francisco Animal Care and Control," "Chattanooga Animal Services" or "Multnomah

County Animal Control." Municipal shelters may or may not have foster programs, although increasing pressure from communities eager to reduce euthanasia numbers have motivated a greater number of these shelters to add fostering to their list of services.

The shelter is part of the municipal "animal control" program, charged with protecting citizens from animals. They are usually responsible for enforcing city or county laws and regulations regarding animals, and may also investigate cases of animal cruelty and offer education programs. Their enforcement staff may be called "animal control officers," "animal services officers," "dog wardens," "dog law officers," or some other such regulatory-sounding name.

"Animal Control" may be its own department in local government, or can be managed by a police department, department of public works, health department, department of parks and recreation, or some other division. Priority of services often depends on which department oversees their work. If under the health department, high priority is placed on "rabies control" efforts; if under the police department, enforcement of animal control laws may take center stage. If you travel up the organizational tree you eventually reach a board of supervisors, a city council, a mayor, or whatever office is at the top of your particular governmental hierarchy.

Some shelters have spacious, clean and inviting facilities and innovative programs. Others, not as much.

Private non-profit shelter. As the name implies, this is a 501(c)3 not-for-profit organization with a board of directors and bylaws that govern the mission and policies of the group. Its mission is to protect animals from people, which often includes a strong educational component. When applying for non-profit status, in most states these agencies are incorporated for the "prevention of cruelty to animals." They may have members, and members may or may not have voting privileges. Most non-profit shelters do have foster programs of some sort.

These groups have names like "Marin Humane Society," "Houston SPCA," "Chicago Anti-Cruelty Society," "Denver Dumb Friends League." Same type of organization: non-profit animal protection agencies—just different names. To emphasize the point, note that "SPCAs" across the country have *no* affiliation with the ASPCA, and your local humane society is *not* a branch of the Humane Society of the United States.

These shelters usually keep animals as long as they can, have active adoption, education and spay/neuter programs, and strive for low euthanasia rates, but can't always succeed, depending on whether they are "open-door" or "limited admission" shelters. They may also be involved in humane investigations, rescues and cruelty case prosecutions. Cruelty enforcement workers are often given titles such as "humane officer" or "cruelty investigator." At the top of the non-profit organizational chart is the president of the board, chair of the board, CEO, or other such title.

Non-profit shelter with an animal control contract. Some non-profit shelters contract with local community government to perform the function of animal control alongside their humane society mission. Under this arrangement, the shelter is still governed by its board of directors, but must respond to the contracting government over issues related to the contracted services.

The contract may be only to *house* stray animals for one or more municipal animal control agencies or it may be to perform field enforcement services as well as sheltering services. These services involve issues such as animals running at large, barking and other "nuisance" complaints, enforcement of licensing and "sanitation" (pooper scooper) laws, etc., and sometimes include enforcement of anti-cruelty laws. Non-profit shelters sometimes take on government contracts for financial reasons—some rely on government dollars to survive—sometimes for humanitarian reasons, in the belief that the non-profit shelter can do a better job of caring for the animals. Sometimes it's both. Because the two missions are in conflict—protecting humans versus protecting animals, this arrangement can have a deleterious effect on community support for the shelter; actions such as issuing citations for leash-law violations, charging a fee for people to reclaim their impounded dogs, or declaring dogs "dangerous or potentially dangerous" don't endear the organization to potential supporters. The issues are often no-win for the shelter—regardless of the action taken, someone is likely to be unhappy.

Non-profit shelters that have government contracts usually euthanize greater numbers of animals, since they are compelled to accept all stray animals as defined by the contract. This group of animals is likely to include some of the least potentially adoptable animals in any given community.

Animal rescue groups. These may or may not be 501(c)3 not-for-profit organizations, and they may or may not be so-called "no-kill." A minority of rescue groups have actual shelters; most make widespread use of foster homes in order to accommodate the numbers of dogs they wish to rescue. Some actually operate "for-profit" rescues.

Breed rescue groups that operate under the auspices of their breed clubs are usually not-for-profit with a governing board of directors, and for the most part are realistic about euthanizing dogs who aren't good adoption prospects—although not always. They tend to use scarce resources wisely, and make thoughtful and difficult decisions about how to help the greatest number of dogs with those limited resources. But not always.

Non-breed-affiliated rescues and mixed-breed rescues can run the gamut from responsible 501(c)3 legitimate non-profit rescues to private adoption agencies to hoarders masking as rescues. We'll discuss hoarders at greater length shortly.

The bottom line is that any of these sources are fertile ground for the would-be foster parent. Whatever the pros and cons of the various shelters, private adoption agencies, rescue groups, or other sources of canine companions, they can all provide dogs who would benefit greatly from a stay at a knowledgeable and capable foster home. Make a list of the groups geographically accessible to you, and then do your homework to find the one(s) with which you will be compatible and able to do the most good for the most dogs.

Doing your homework

Doing your homework means doing a little investigating. Start by contacting each of the organizations on your list and ask if they are accepting foster home volunteers for their dogs. If the answer is yes, ask what kind of volunteer orientation and training is required. If they say none, move them to the bottom of your list. A responsible organization will make sure their volunteers have had training in understanding canine body language and behavior, and are given information about the agency they are working with. A group that fails to provide training may well cut other critically important corners as well and may be putting the welfare of their animals at risk by doing so.

Ready for a field trip? Some of the groups on your list will have physical shelters. Visit the ones that do, even if it hurts your heart to see dogs in shelter kennels. You need to know that any group you choose to work with takes proper care of the dogs for whom they are responsible. Kennels may be crowded, but they should be reasonably clean, and you should not see obviously sick, debilitated or injured dogs mixed in with the general population. As much as you may feel drawn to offer to help dogs who are in a substandard facility, it will be a nightmare to try to foster sick dogs, and you will likely be a victim of burnout before you can help very many of them. To say nothing of the risk to your *own* dogs at home if you take a chance on bringing home sick dogs. Of course, any shelter can have dogs who are ill, but in better shelters sick dogs are kept in isolation and treated, disease risk is controlled and the threat to your own dogs is minimized.

After you have visited the facilities in your area, move the good ones to the top of your list and the bad ones to the bottom. If you have seen dogs who are truly suffering from neglect, seriously consider reporting them to the authorities—that's the best and fastest way you can ensure those dogs get the help they need.

By now you should have a select few at the top of your list that are candidates for your fostering services. Contact them again, and ask to meet with the person in charge of their foster program. The best groups will want to screen you as much as you want to screen them; be prepared to go through an application and interview process. Have your own list of interview questions in hand when you go; they will help you decide if this is a group you want to work with or not.

Ten sample interview questions to ask your potential fostering organization:

1. **What is your organizational structure?** You want to know if they are a municipal shelter, a humane society (SPCA, etc.) a breed rescue, or all-breed/mixed-breed rescue. You will also want to know if they are a 501(c)3 organization, in which case your fostering expenses (including travel) are tax deductible—but not the time that you volunteer. If they are municipal, what governmental body oversees their work? If they are 501(c)3 non-profit, who is their board of directors?

2. **What is your organizational mission/philosophy?** Good to know if they are able to articulate this—and to know if it's something you are comfortable with.

3. **How many animals do you take in per year? Where do they come from? How many do you euthanize, and what guidelines/policies do you follow for euthanasia?** If they tell you they don't *ever* euthanize, they are either not telling the truth, or they are keeping dogs alive without consideration for quality of life. Even if they are a limited intake group and technically don't do euthanasia themselves because they take dogs to another group or a veterinarian for the procedure, they should still acknowledge that some of their dogs *are* euthanized. *All* legitimate groups euthanize sometimes, for health or behavioral reasons, even the "no-kill" groups. You want to work with a group that is honest, ethical and humane, and you want to know under what circumstances they do choose to euthanize. You can then decide your own comfort level. Some fantastic foster homes are able to deal with the dissonance of loving dogs and fostering for groups that have higher euthanasia rates; others are not. In fact, some choose to work with full-service shelters *because* those dogs may be at greater risk for euthanasia. There's no right answer here; dogs in all kinds of shelters and rescues benefit from foster programs.

4. **How much disease do you encounter?** This is more often an issue for shelters that house large numbers of animals at a single location than it is for rescues that have their dogs in foster care, but rescue dogs can get sick too. Does the group vaccinate dogs on intake? This is the standard practice in the industry, and is the best approach to minimizing disease transmission. Upper respiratory infection is the most common, but the more serious, possibly fatal diseases such as parvovirus and distemper also occur. Bringing foster dogs home means potentially exposing your own dogs to these germs, and risking that *they* may become ill—perhaps seriously so. The more illness at the shelter, the greater the risk to your dog. Are you willing to take that risk? Who pays for treatment if *your* dogs become ill?

5. **How do you euthanize?** You don't need all the details, you just want to be sure they euthanize by intravenous injection (into a vein, *not* directly into the heart), not by carbon monoxide or decompression chamber, shooting, or worse.

6. **Do you do behavior assessments?** Good shelters and rescues use some kind of standardized procedure to evaluate their dogs, both to identify behaviors that might need to be modified (or that may identify the dog as too risky for adoption), as well as to improve the adoption process by helping to match dogs with appropriate homes. Ask if you can observe assessments being performed, and see if you are comfortable with what they do. If they don't do assessments, consider carefully whether this is a group you want to be involved with, whether you want to take on the additional risks of fostering a dog who hasn't been evaluated. Even with evaluations, you will inevitably discover additional behaviors when you bring the dog into your home—desirable ones as well as undesirable ones.

7. **Who pays expenses?** Does the organization you are interviewing provide food and veterinary care?

Do they reimburse you for those? Or are they expecting you to pay for the care of your foster dog out of your own pocket? If they put a cap on veterinary expenses, will they allow you to pay for costs above and beyond the cap? If the dog needs to work with a training/behavior professional, will they pay for that? If you go away on vacation, will they pay for boarding or a petsitter?

8. **Are you insured?** And perhaps more on point—does their insurance cover you? You may think you're not worried about this—until your foster dog bites someone and you get sued. If you lose your home in a lawsuit, you won't be able to help future foster dogs. Legitimate organizations have insurance to cover their volunteers.

9. **What are your rules, restrictions, procedures and policies?** Can you take your foster dogs with you in public? Travelling? Hiking? Camping? Can they go to dog parks? *Should* they go to dog parks? Are there things the dogs can or cannot do in your home? What happens if your foster dog bites someone? How long is a foster dog likely to stay in a foster home? Are they vaccinated before you take them home? (They should, at the very minimum, have a rabies shot before fosters are released into your care.) Do they need dog licenses? If so, who pays for that?

10. **How does the adoption process work?** Are you, the foster parent, allowed to be a part of that process? (The answer should be "yes.") Are potential adopters carefully screened? Do you get veto power if you think the home is not appropriate for your foster? Are the dogs adopted directly from your home, or do they have to go back to a shelter or kennel? Are you expected/allowed to bring your fosters to meet-and-greet events? What if *you* decide you want to adopt one or more of your fosters—is that allowed?

Ask all the questions you need to in order to make sure you are comfortable with the mission, policies, procedures and ethics of the organization for whom you choose to foster.

You likely have some additional questions of your own. Good for you! Regardless of how many questions you want to ask, the person you are meeting with should be willing to patiently answer them all. There aren't necessarily right and wrong answers—you are simply seeking information to determine if you are comfortable partnering with this particular organization, or if you should keep looking.

Making appropriate foster selections

You've chosen the organization(s) you want to work with. The rubber is about to meet the road—you're actually going to bring home your first foster dog. How do you make wise decisions, so that your fostering experiences are positive ones for you, will provide long-term benefits for your foster choices, and won't have a negative impact on your household and its other occupants, human or otherwise? Here are some tips:

1. **Start easy.** As a novice foster parent, you have every right to ask your shelter or rescue group to start you off with a low-maintenance foster dog. Perhaps an underage puppy (or pair of puppies) with no health or behavior issues, who just needs

to grow up for a couple of weeks before she can be responsibly placed. Puppies should generally be adopted to new homes at age eight weeks or older. In some states, Maryland for one, it's illegal to place them prior to eight weeks. You'll need to do some good socialization work during the time you have them, but nothing heroic. At five to six weeks they can eat on their own, so if you foster pups from five weeks to eight weeks you don't need to bottle feed, just give them plenty of good social exposure, and keep them safe and well fed. "Good social exposure" means positive interactions— the puppy enjoys the experiences and learns that the world is a safe and happy place. Good social exposures turn your pup into a brave and confident optimist. "Bad" social exposures, in which the pup becomes frightened or unhappy, teach him the world is a scary place, and every new thing he encounters will cause him to be more fearful and timid. An unsocialized or badly socialized pup grows up to be a pessimist. Another "easy" might be an otherwise mentally and physically healthy dog recovering from an injury or surgery—broken leg in a cast, umbilical hernia repair, cherry eye, etc. They'll need to be on "restricted activity"—no romping with the home dogs—but don't require Clara Barton-level health care or extensive behavior modification.

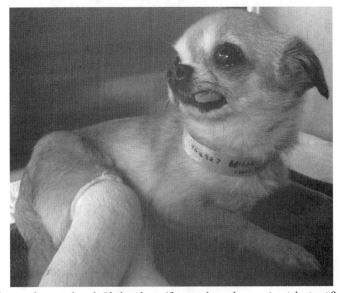

This undersocialized Chihuahua (from a hoarder case) with significant medical issues is not a good choice for the novice fosterer.

2. **Do a pre-foster meeting.** All human family members *must* meet the dog before you commit to fostering this particular one. You don't want to go by yourself to choose a foster, only to bring him home and discover he doesn't like big men with glasses and beards—because of course your husband is a big man with glasses and a beard. Or the dog is not good with small children, and your family includes four children, ages three, five, six and seven.

If you have small children, your foster dogs must love kids!

3. **Consider your own dogs' preferences.** If your dogs get along better with males, foster male dogs. If your small dogs are stressed by big dogs, foster small dogs. If your big dogs are too rough for small dogs, foster big dogs. You get the picture. Think it through.

4. **Consider your own preferences and those of your human housemates.** Acknowledge your own biases, if you have them. If you turn up your nose at frou-frou dogs, foster "manly" dogs. If you shudder at canine drool, stay away from St. Bernards, Bloodhounds and other slobbery breeds. If you don't want light-colored dog hair all over your dark-colored décor, foster dark-colored dogs. If you have a child who is fearful of large dogs, foster smaller dogs. If your spouse has a "thing" against smoosh-faced dogs, avoid fostering brachycephalic breeds like Pugs and Boxers. There are plenty of dogs of all kinds who need fostering; you can work around just about anything.

5. **Understand your limitations.** If you have to be gone all day, or can't tolerate an interrupted sleep cycle, don't volunteer to foster neonates (very young puppies) who need to be bottle fed every few hours. If you have physical limitations, avoid rambunctious fosters with high exercise needs. Small dogs and low-key senior dogs might be a better choice for you.

6. **Capitalize on your strengths.** If you have a foundation of behavior and training knowledge—or are willing to learn—you can take on fosters who need to learn good manners or must have inappropriate behaviors modified in order to maximize their adoption potential. Helping with successful placement of these dogs can be exceptionally rewarding—but often takes an additional investment of time, energy and emotion (and perhaps money to pay for training classes or behavior consults, if your organization doesn't offer such resources). If you love to exercise, look for high-energy types (Herding and Sporting breeds, and Sighthounds). If you're more of a couch potato, you might specialize in fostering senior dogs, or less active breeds like English Bulldogs, Newfoundlands and Pugs.

Setting limits

Here's the major caveat about fostering. It's really easy to get sucked into the "just one more" mentality, and before you know it you're a hoarder, with dogs stashed in every nook and cranny of your residence. The smell of dog feces begins to permeate your carpeting and your clothing. People move away from you in public. Friends and family stop visiting, your children spend more and more time at their friends' homes, and your spouse threatens to leave you.

Okay, so maybe that's a bit of an exaggeration (although it really *does* happen!), but even to a lesser degree, it's important that you not let your good intentions and commitment to fostering degrade the quality of your own life or the life of your family members, including your canine family members. Don't let your shelter or rescue group—or your own heart—guilt-trip you into taking on more than you can reasonably handle. It's a fact that since the inception of the so-called "no-kill" movement, the numbers of hoarder cases have skyrocketed. What used to be the few-times-a-year sensational "cat lady" case has turned into weekly new reports of cases where authorities remove massive numbers of dogs, cats and other animals (100-plus) from the homes of hoarders, some of whom started out as well-intentioned rescuers and foster homes.

I have, on multiple occasions, had clients come for private consultations because their dogs have developed serious behavioral issues as a result of a constant parade of foster dogs in and out of their world. I recently had a client whose foster dog was being attacked by her own dogs (she had eight of her own). Not fair to the foster dog to put her through that stressful experience, and not fair to her own dogs, who were being stressed by a constantly changing canine population.

It also doesn't do your fosters, your family or your own dogs any good if you stretch your resources to the breaking point. If you're expected to proved food and veterinary care for your fosters and that interferes with your ability to provide for the needs of your own family, you need to rethink the plan.

Good intentions alone aren't good enough. If your environment isn't conducive to fostering success, for whatever reason, you can explore other ways to help homeless dogs in your community. Volunteer to walk dogs at your local shelter. Offer to assist with fundraisers for your favorite animal organization. Take dogs to the local television station for media appearances. Drive dogs to and from vet hospital appointments. Transport dogs to meet-and-greet events, and talk with potential adopters there. Sign up to do home visits as part of your agency's adoption screening process. Ask your group how else you might be able to help. There are lots of ways to do good for the homeless dogs that you so badly want to help, if fostering isn't a realistic choice for you.

If your circumstances aren't conducive to fostering success, there are plenty of other volunteer opportunities to help homeless animals.

3

Bringing Your Foster Home

Pre-foster preparations

Assuming you've already talked your entire family into supporting your foster project, what else do you need to do to get ready for your first foster? Look around your house. If you have resident dogs you've likely already done some dog-proofing, but you may need to do a lot more because your new foster dog will probably prove to be more challenging.

Dogs end up in shelters and rescues because their relationship with their former humans was somehow broken. Sometimes this occurs through some tragedy that separated them from a loving home. However, sometimes it happens because the dogs weren't trained or well managed and were allowed to practice behaviors that will challenge even the most experienced dog-keeper. Management measures that worked with your personal dogs may not be adequate for your fosters. Be prepared to invest in crates, baby gates, tethers, exercise pens, and perhaps even a chain link kennel, if necessary, to keep your foster charges out of trouble and to set them up for behavioral success.

Are your crates, baby gates and tethers in place?

Remember how to puppy-proof? Even if your foster isn't a puppy, all things chewable need to be picked up and put away—shoes, clothes, toys, valuable keepsakes and more. Electrical cords should be safely protected. No leaving food on tables and counters. Back to covered garbage cans, or trash receptacles in closets. You might even consider installing baby-proof latches on cupboards, in case you luck out with a foster dog who is exceptionally gifted. Time to invest in a supply of interactive toys and indestructible chewies—Kongs™ to stuff, tough squeaker toys, and more—to keep your foster dog productively occupied.

All this needs to be done *before* your first foster dog comes home. If you wait to install management measures until *after* the need for them arises, you've given your canine visitor yet one more opportunity to be rewarded for undesirable behaviors, making it that much harder to successfully integrate him into your home, and more importantly, his next, and hopefully forever home. Your goal is to set each foster dog up for success from the moment he arrives.

Speaking of setting up for success—if your partner organization has not already done so, you'll want to make an appointment to stop by your veterinarian's office (or theirs) with your foster dog on the way home from picking him up. You want to know that he has been vaccinated, checked for parasites (and treated, if necessary) and that he has a clean bill of health. A veterinarian can check for visible signs of disease (mange, ringworm, ocular or nasal discharge, fever) and talk to you about whether she recommends quarantining your new charge—and if so, for how long—to protect your own dogs from any transmissible diseases that your foster may bring home. While you're at it, make sure your own dogs are up-to-date on all *their* preventative health care, to decrease the chances that their health will suffer from your foster care endeavors.

Foster dog history

It's tempting, when you don't have information about your foster dog's history, to make stuff up based on what you see. This is a mistake that is commonly made, and it's one that makes caring for and training your foster dog more difficult. Don't do it. You will probably be wrong, and by providing your dog's partner organization and future owners with misinformation, you could be creating problems where none exist. Many people think a fearful dog has been abused—and if the dog is fearful of large bearded men, the fairy tale is that he must have been abused by one or more large bearded men. What is more probable is that he simply wasn't well socialized, and because large bearded men are inherently more scary-looking, the foster reacts more fearfully to them.

These days, many of dogs of unknown history who have scars on their bodies are assumed to have been "bait dogs." The fantasy is that there are scads of reprehensible humans with fighting dogs grabbing up every stray dog they see to use as bait for their Pit Bulls. It's really not so. For one thing, the average dog, if he *were* used as a bait dog, probably wouldn't survive. There are lots of reasons your foster dog may have scars, especially if he had to fend for himself for any length of time. Don't go telling people he was a bait dog unless you know for a fact he was rescued as part of a raid on a dog fighting operation, or otherwise came directly from an actual dogfighter's hands. When you're reporting information to your organization or your dog's next and final home, just stick to the known facts.

Be prepared

You've already made the necessary advance preparations. Now the day you've been waiting for has finally arrived. Excitement is high—your foster dog is coming home! What can you do to make this transition as easy as possible for everyone concerned?

First, double check that you have all your equipment on hand—crates, tethers, exercise pens and baby gates, and that you have a supply of the food your foster has been eating. A sudden food-switch can trigger a nasty bout of gastrointestinal upset and resulting diarrhea—not something you need to be dealing with along with all the other chaos that can accompany the arrival of a new dog in your household. Stock your freezer with enough stuffed Kongs and/or other food toys to keep your foster busy for several days. (You'll want to be sure to replace them as they are emptied, so you don't run out.)

Successful introductions

If your foster has a clean bill of health and you don't need to quarantine, one of your first orders of business may be to introduce him to your own dogs. In preparation for that, make sure your dogs are well exercised and on the tired side. Whether someone brings your foster to you, or you go pick him up yourself, give him a chance to stretch his legs, go to the bathroom and recover from the stress of the trip before you try dog-dog introductions.

There are a number of factors to keep in mind that can increase the likelihood of a positive outcome when introducing your foster dog into your home. A peaceful first introduction sets the stage for fostering success. The more heavily you can weigh the odds in your favor for that first encounter, the greater your chance for peace in the pack. The factors to keep in mind include:

- Timing
- Location
- Number of skilled handlers
- Knowing and understanding—to the greatest extent possible—the personalities and histories of all the dogs involved
- Introduction process

Dog-dog introductions need to be carefully done to set yourself up for success.

Timing. It's best to introduce your foster dog to your home when things are otherwise calm and reasonably stress-free. You want to allow ample time for a leisurely introduction process and a low-key adjustment period with adequate supervision. In addition, you need to be able to iron out any wrinkles that may appear. This may mean taking time off work, in case the dogs don't hit it off instantly. Holidays are generally *not* the ideal time for introductions unless, for you, "home for the holidays," means lots of quiet time spent alone with your fur-family.

Of course, you can't always control the timing. A shelter or rescue dog may be facing a ticking clock that dictates a speedy transition to foster. Just do the best you can to arrange for your foster to arrive at a time that optimizes your success potential. If he has to come at an inopportune time, arrange to keep him in a room or building isolated from your own dogs until proper introductions can be made.

Location, location, location. It's best to introduce dogs in neutral territory—ideally outdoors, in a large, open, safely fenced space. The more trapped a dog feels, the more his stress will push him toward defensive aggression. Plus, when you do introductions in one dog's territory, it gives that dog the home-field advantage, and you risk displays of territorial aggression.

Optimum options include a fenced yard other than your own, an off-leash dog park at low-use time (that is, with *no* other dogs present), a tennis court (caution—many tennis courts understandably prohibit dogs), or a large, open, uncluttered indoor area such as someone's unfinished basement. If none of those are available to you, your own fenced yard may be your best choice.

Number of skilled handlers. Ideally, you'll want at least one handler per dog. One *skilled* handler, that is—a person who is comfortable handling dogs, and who you trust to follow your instructions or, alternatively, a qualified positive behavior professional who can coach *you* on the process. Someone who panics and intervenes unnecessarily, or who thinks they know it all and that rough "dominance" handling is the right approach, can botch the whole job by adding stress to dogs who are still sorting out relationships. If you don't have access to skilled handlers, at least find handlers who are good at following instructions and don't succumb easily to hysterical behavior. If you can't find those, you're better off with fewer handlers, although you should have at least one other person present, if for no other reason than to call 911 if the situation gets out of hand.

Knowing and understanding the personalities and histories of the dogs. You may not know much about your incoming foster, especially if he was a stray, or seized in a neglect or cruelty case. When dogs are surrendered by their owners, a good organization takes a through history in order to facilitate interim care and appropriate rehoming. You should, however, have a pretty good sense of your own dogs' canine social skills. Do they play well with others at the dog park? During playtime at good manners class? With their own packmates? How do they act with doggie visitors to their home? During chance encounters with other canines on the streets?

If you have reason to believe that your dogs are anything less than gregarious with **conspecifics** (others of their own species) due to a history of aggressive behavior with other dogs, or if you just aren't confident about refereeing the introductions yourself, you might do well to rethink your fostering project, or at least engage the services of a qualified behavior professional. She will be able to help you read and understand your dogs' body language, and optimize the potential for success.

Hopefully, you've already given great consideration to good personality matches when you selected your foster dog. If you have a dog in your pack who likes to assert himself, you're wise to choose a new dog who's happy to maintain a lower profile in the hierarchy. If your current dog is a shrinking violet, she'll be happiest with a new companion who doesn't bully her mercilessly. If you have one of those canine gems who gets along with everyone, then you have more fostering options. If you want your own dog to be able to be "top dog," then look for soft, appeasing-type fosters. If you don't care where your easygoing dog ends up in the hierarchy, then you have the entire canine personality continuum to choose from.

The introduction process

You are safest introducing your new foster to your easier dogs first, one at a time. Assuming all goes well with the one-on-ones, then try a threesome, adding additional dogs as behavior allows.

The process I've found most successful is to start with dogs on leashes on opposite sides of the enclosed space. Try to keep leashes loose, if possible. Watch the dogs' behavior. They should seem interested in each other, alert without excessive arousal. Ideally you'll see tails wagging at half-mast, soft, wriggling body postures, play bows, ears back, squinty eyes, no direct eye contact. These are clear expressions of non-aggressive social invitation.

Warning signs include stiffness in the body, standing tall, ears pricked hard forward, growling, hard direct eye contact, stiffly-raised fast wagging tails, perhaps even lunging on the leash and aggressive barking.

There is tension in this interaction—be careful!

If you see reassuring social behavior, proceed with the approach until the dogs are about ten feet apart. If they continue to show unambiguous signs of friendliness, drop the leashes and let them meet. I prefer not to let dogs meet and greet with handlers holding the leashes. Leashes tend to interfere with the dogs' ability to greet normally, and can actually induce dogs to give false body language signals, resulting in aggression. For example, a tight leash can stiffen and raise a dog's front end, causing him to look more tense and offensive than he means to be, which in turn can cause the other dog to react offensively. A defensive dog who wants to retreat may feel trapped because of the leash, and act aggressively because he can't move away.

Leave leashes on the dogs initially, dragging freely on the ground, so you can grab them and separate dogs easily if necessary. Monitor the greeting. You are likely to see some normal jockeying for position and some tension, as they sniff and circle, and then erupt into play. As soon as you can tell that they're getting along, remove leashes and let them play unencumbered. Watch that the play doesn't escalate into excessive arousal (which can lead to aggression), but remember that it's normal and acceptable for dogs to growl and bite each other in play. As long as both dogs are enjoying the action, it's a good thing.

If, however, you see warning signs as you approach with the dogs on leash, you'll need to go more slowly. Most commonly you'll see behavior somewhere on the continuum between completely relaxed and friendly and outright aggression. You'll need to make a judgment call about whether the intensity of the behavior is such that you need to stop and seek professional assistance, or low enough that you can proceed with caution.

If you do decide to proceed, interrupt prolonged hard eye contact by having each handler divert her dog's attention with bits of tasty treats while at a safe distance. Continue to work with the dogs in each other's presence, watching for signs of decreasing arousal. Walk around the available space with the dogs at maximum distance, gradually bringing them closer together until they are walking parallel to each other.

It's important that you stay calm and relaxed during this process. If you jerk or tighten the leash or yell at the dogs, you'll add stress to the situation and make it harder for them to relax.

When you see clear signs that the dogs have relaxed with each other, you may decide to proceed with dropped-leash greetings, or you may choose to end the introduction for the time being, and do several more on-leash sessions over a period of several days before dropping leashes. This is where your experience and instincts come into play. It's better to err on the side of caution, and do several more on-leash sessions to make sure the dogs are comfortable with each other. Meanwhile, you'll need to manage the dogs so they don't have free access to each other. If you're not confident in your judgment about body language, you may choose to enlist the help of a professional at this point in the process.

Dogs playing happily together are a joy to behold.

If tensions between the dogs escalate or maintain at the same level of intensity despite your on-leash work over several sessions, the wise choice may be to look for a different dog to foster, or commit to diligently keeping the dogs separated while the foster is with you. *(Note: this can be very stressful for all concerned, human and canine.)* Alternatively, you may want to do ongoing work with a behavior professional to try to make the relationship work, knowing that management may be a large part of your life for the foreseeable future.

Be careful if you see *no* interaction between the two dogs you're trying to introduce. What appears to be calm acceptance of each other may in fact be avoidance behavior—neither dog is comfortable with the other, and they choose to deal with it by not dealing with it. The problem with this is that sooner or later the dogs *will* interact if they're both living in your home, and the discomfort may well develop into aggression. You really want to see *some* interaction between dogs in order to make a decision about how to manage and live with your foster dog.

After introductions are over and the dogs are hanging out happily together you can breathe, rest and get on with life. Regardless of how well they seem to get along, however, it is wise to separate your foster from the others when you need to leave them alone, for at least a few months, and unless and until you are 100 percent confident they are completely comfortable and compatible with each other. Intrapack behavior will change over the first few months and tensions will wax and wane as relationships get sorted out and your foster gains confidence. You can easily prevent disaster by separating dogs when no humans are home to supervise.

More tips for successful introductions

Here are some additional things you can do to increase your potential for successful introductions:

1. Exercise the dogs before initiating introductions. Happily tired dogs are more likely to interact well than those who are bursting with energy.

2. Have tools within easy reach in case you need to interrupt an aggressive interaction.

3. Be sure to remove toys and other high-value chew objects from the introduction area to minimize potential for guarding incidents.

4. Use extra caution when introducing a puppy to adult dogs. A bad experience with an aggressive dog can have a significant negative influence on a pup's future social behavior.

5. Use extra caution when introducing a new dog to senior members of your pack, especially if the new dog is an adolescent or a puppy. Your geriatric dogs shouldn't have to defend themselves from overwhelming attentions from fractious youngsters. Senior dogs often have aches and pains that can be exacerbated by pouncing puppies—and cause an aggressive response from the oldster as he tries to protect himself. Be prepared to implement management tools to protect your seniors from the young'uns.

6. Consider size. Jean Donaldson, past director of the San Francisco SPCA's Academy for Dog Trainers, recommends no more than a 25-pound difference in size between dogs in a household or playgroup. More than that, she warns, and you risk **predatory drift**, where the larger dog's brain suddenly perceives a small running dog as a prey object such as a bunny

or squirrel, and shifts from play to food-acquisition mode, sometimes with tragic results. Know that if you choose to introduce a new dog to a situation where there is a large size disparity you may be taking additional risks with your dogs' safety during introductions and thereafter.

Introductions to other species

Many animal lovers have multiple species in their homes. You will want to use caution when introducing your foster to other non-human family members, not just to other dogs. Watch for manifestations of strong predatory behavior in your foster:

- Long, hard stares at cats, birds, hamsters, livestock, etc.

- Over-arousal in the presence of other species—lunging, barking, yelping, jumping.

- Chasing other animals—small animals can be injured or killed, large animals (horses) can injure or kill your foster.

- Persistent attempts to get into rooms/areas where other animals are kept.

If you see any of these, you will need extra management and security to keep your other animals safe, such as latches and padlocks or slide bolts on doors, and a house rule that the foster is never allowed to be loose without direct supervision. You may want to reconsider this foster choice and exchange him for another, for the safety of your other family members. Even babies/small children can be perceived as prey by dogs who have very strong predatory behaviors.

Assuming an absence of strong predatory behavior, you will still want to proceed cautiously with your introductions. This is an excellent time for a round (or three) of counter conditioning. (See Chapter 6 for a discussion of counter conditioning.) Each time you present your "other species"—cat, for example—have your foster on leash and feed him bits of boiled, baked or canned chicken. Let him look at the cat. Feed. Look at the cat. Feed. Over and over again, until every time he sees a cat he immediately looks to you for chicken. You are creating a classical association: "cat equals chicken" rather than letting him create his own: "cat equals fun game of chase." Again, until you are 100 percent positive your other animals are safe, *never* leave your foster dog free to discover that other housemates are fun to chase and perhaps kill.

When it doesn't go well—breaking up a fight

I hope you will never need these! Still, it doesn't hurt to have some of them on hand. Please note: these work because they are aversive and are only to be used in a crisis. I am not in *any way* recommending them as regular training tools or methods.

Hands-off intervention

- Blasting dogs with water from a nearby hose may work, assuming a hose happens to be nearby with a powerful enough spray. A good tool to keep in your arsenal for the right time and place—not particularly useful, however, when there's no hose handy!

- One of the easily portable aversive sprays, such as Halt!™ (dog repellent spray) might be an effective alternative to the hose. There are laws in some jurisdictions requiring that users of pepper spray products complete a training course and carry a permit. You could also talk to your veterinarian about a recipe for a home-made spray that could work as a deterrent but not be permanently harmful to dogs.

- In a pinch, even a fire extinguisher might just happen to be a handy and effective aversive tool.

- Some doggie daycare providers swear by marine air horns. Available at boating supply stores, air horns can be effective at interrupting fighting dogs. You also risk damage to eardrums, both canine and human, and take a chance of frightening your own dog beyond repair.

Physical intervention with objects

- Attach a couple of handles to a sheet of plywood so you can lower it between two sparring dogs.

- Animal care and control professionals use a tool called a "control stick" or "catch pole." This is a long metal pole with a noose at the end. This tool can be of use in separating fighting dogs. Anyone can purchase one of these—but if you do, ask a professional to teach you how to use one properly and humanely.

- People who train dogs to fight—and some Pit Bull owners who don't fight their dogs but know the breed's potential—always carry a **parting stick** or **breaking stick** with them. This is usually a carved wooden hammer handle, tapered to a rounded point at one end. When two dogs are locked in combat, the parting stick can be forced between a dog's teeth and turned sideways to pry open the jaws. Parting sticks can break a dog's teeth, and a dog whose jaws have just been "parted" may turn on the person doing the parting. Like many other techniques offered here, this method should only be considered for dire emergencies. (Breaking sticks can be found online at: pbrc. net/shop/bsticks.html.)

- A blanket can also be useful. Tossed over the fighters, one over each, blankets muffle outside stimuli, reducing arousal. This also allows the humans to physically separate the combatants by picking up the wrapped pooches with less risk of a serious bite—the blanket will cushion the effect of teeth on skin if the dog does whirl and bite.

- When a dog's life and limb are at stake, extreme measures may be called for. You can wrap a leash round the aggressor's neck or get hold of a collar and twist to cut off the dog's airflow, until he lets go to try to get a breath of air, then pull the dogs apart. This could be more difficult than it sounds. It might be difficult to get a leash around the neck of a dog who is "attached" by the mouth to another dog without getting your hands in harm's way; grabbing a collar to twist also puts hands in close proximity to teeth.

Physical intervention by humans
- One rather drastic technique was observed at a dog show some 20 years ago. Two dogs got into it and were going to cause major damage. The elderly judge was a tiny woman, and she had the handlers both grab their dogs and hold on *tight*. Then she went up and took the dog on top by the tail and inserted her thumb into his rectum. He let go in an instant and whirled around to see what was happening. The judge excused the two dogs, calmly washed her hands, and continued her classes without a hitch—just as if it happened every day.

- Another approach could be difficult if the aggressor is a 150-pound St. Bernard, but may be worth trying with a smaller dog in a one-on-one fight. It is not recommended for a multi-dog brawl. Lift the rear of the clearly identified aggressor so that he is suspended with his forefeet barely touching the ground. The dog lets go, and the target can scoot free. Alternatively, humans can grab *both* dogs and pull them apart. Supposedly, in this position the dog is not able to turn on the human suspending him, although I'm not giving any guarantees.

Armed and ready

Now, all you need to do is stuff a canister of Halt in your pocket, attach a parting stick to your belt, carry a blanket over your arm, prop a control stick in the corner, balance a sheet of plywood on your head, wear an air horn around your neck, be sure you have at least two friends with you to hold dogs while you put your thumb in private places, and you're ready for anything.

Seriously, if and when that fight happens, take a deep breath, resist your instincts to yell or leap in the middle of the fray, quickly review your available options, and choose the one—or ones—that are most likely to work in that place and time. When the fight is over and no one is being rushed to the hospital in an ambulance, remember to take a moment to relax and breathe, and then congratulate yourself for your quick thinking.

Adjustment period

Let's hope your introductions went well. Whether they did, and you have relative harmony in your home, or they didn't, and you're juggling management for the duration, your foster dog will go through an adjustment period as he settles into his new digs. He, of course, doesn't know these are temporary arrangements for him, although with all he's been through in recent weeks or months he's likely to be pretty stressed, with no expectations that this lodging will be any more permanent for him than the last.

Change is stressful for all of us. Some dogs handle change and stress better than others, but each time your foster goes through the stress of a change it makes it more difficult for him to adapt. We all want to have control of our worlds, and he has lost all control of his—from the time he perhaps got lost or abandoned to the scary streets of your city, was picked up and taken to the shelter where he was surrounded by barking dogs, banging kennel doors and a constant parade of strange humans, through exams, assessments and other handling procedures, to now—a new home, with new dogs, new humans and new rules.

Many rescuers suggest that the first month or two with any new dog is a "honeymoon period," in which the dog's behaviors may be somewhat inhibited or suppressed. That may be true, and it may be more of a "trial period," where your dog is trying out a variety of behavior strategies, and selecting the ones that seem to be working for him—those that are either positively reinforced (they make good stuff happen) or negatively reinforced (they make bad stuff go away). This holds true for foster dogs as well as it does for newly adopted dogs. Your foster will go through this period again when he is adopted to his forever home.

Over time, the successful strategies will increase in frequency and perhaps intensity. By the end of the month, if you haven't carefully managed your foster to be sure the behaviors you *like* are the ones that are getting reinforced, you may see the increasingly noticeable presence of *undesirable* behaviors. What rescuers refer to as seeing the "real dog" after the first month may just be those behaviors becoming established over time through reinforcement.

You can help your foster charge adjust more easily by creating consistency and structure in his world—rules to live by that everyone who interacts with him follows. You might have different rules for your fosters than you do for your own dogs. Perhaps you allow your dog on the furniture because you enjoy his close companionship and warmth. You allow him to jump on you because you appreciate a happy greeting. That's fine if that's what you like. My own dogs are allowed on the furniture, and some of them sleep in our bed. However, if you set stricter rules for your foster, he won't have to learn stricter rules and face a difficult adjustment period if he is adopted into a home that *doesn't* allow dogs on the furniture and *doesn't* appreciate muddy paws on their Brooks Brothers suits when they arrive home from work.

Of course, the last thing your foster dog needs is the added stress of aversive training methods that use pain and coercion to teach house rules, so be sure you *gently* introduce and reinforce the new structure for him. Rather than kneeing him in the chest when he jumps up on you, for example, turn away from him—and pay attention to him when he has four paws on the floor, or better yet, when he is sitting. Consistently reward him for doing what you want, and set up his environment to make it difficult for him to do what you don't want.

You *can* train your foster dog during the adjustment period and the weeks and months that follow. In fact, any time the two of you are together he is learning, so you might as well take advantage of that and help him learn the *right* things from the very beginning. Arm yourself with some good books on positive reinforcement-based training, and help your foster learn that the world is a safe, consistent, predictable and happy place. He may take some convincing after all he's been through. Or he may come to you already an optimist, and be happy to have you confirm his positive world view.

You will learn a lot about your foster during the time he is with you. The adjustment period will be a time of discovery for both of you. The better you are at documenting what you learn about him and can pass on to his new caretakers, the easier the transition will be to his lifelong home when he leaves your hands. Is he housetrained? How much easier for him if you can tell his new humans what his "I have to go out" signals are. Chances are *you* will have to learn them by trial and error. Is he afraid of thunder? The shelter probably didn't know that, but now you do—and you can provide his adopters with good information on how to live with, and help, a thunder-phobic dog. He may have come to you with no history of interactions with children, but you see how warm, soft and wriggly he gets when he sees the neighbor's toddler, and now you know that he could probably be placed in a home with small children. You realize he is fiercely dedicated to chasing cats—so you can cross cat owners off his list of prospective homes. You want his next residence change to be his last—and the more information you can share about him the better his chances are of finding the lifelong loving home that every dog deserves.

Socializing your foster dog

If you are caring for one or more foster puppies, you *must* understand the importance of and make a *total commitment* to socialization. **Socialization** means giving your pup *positive* exposures to the world while he's young enough to be forming his world view. *Early socialization is your best immunization against fear and other related behavior problems*—it creates an optimistic puppy who assumes the world is a safe and happy place unless and until proven otherwise. The *most* important period for socializing your foster is from the age of three weeks to fourteen weeks. After that, the window of opportunity starts to close, and it closes quickly. Dogs who miss their early socialization never totally recover, and are much more likely to be pessimists who fall victim to generalized fearfulness which itself can lead to aggression and even medical problems. Ongoing socialization is important throughout a dog's life, but it is nearly impossible to completely repair damage done by failure to socialize during the three to fourteen week period.

If you are fostering puppies, you must be prepared to make the commitment and take the time to socialize them well and properly. That means lots of different positive exposure to and experiences with a wide variety of sights and sounds. No exceptions, no excuses. Photo: Greg Armes

Lots of early socialization will help your foster puppies grow into confident, friendly, outgoing canine companions. They need to meet lots of different people, see lots of different sights, walk on lots of different surfaces and hear lots of different sounds. Failure to socialize creates shy, fearful dogs who can never enjoy the world, and are at high risk for eventually biting someone. Dogs who bite tend to have short, unhappy lives. I know I've repeated this several times, but it is very important. Foster dogs who are stressed by the changes and uncertainty of their life circumstances, and who don't have one or more humans of their own who have made a lifetime commitment to them, are especially at risk.

A dog's adult personality comes from nature (genetics) *and* nurture (environmental influences). If your foster pup is genetically confident, he'll still need *some* socialization to become a well-adjusted adult. If he's genetically shy or timid, he'll need *tons* of socialization to become a normal dog. Because you probably won't know about the genetic makeup of your foster pups, the best insurance against fearful, potential biting adult dogs is to socialize the heck out of *all* puppies.

Note the importance of *positive* experiences. Protect your foster babies from painful or frightening experiences. Don't expose them to excessively loud noises or extreme visual stimuli—like your town's 4th of July fireworks display. Supervise interactions with children so they can't tease and torment your foster babies or encourage inappropriate puppy biting and chasing. Instead, have them feed tasty treats for sitting politely—kids love to learn how to train dogs! Make puppy experiences like trips to the vet for vaccinations as positive as possible—lots of treats in the waiting room, lots of treats while the pups gets their shots, lots of treats from the vet and clinic staff. In fact, make it a point to visit your vet's waiting room with your foster pups when you don't have an appointment, so going to the hospital doesn't always mean being poked and prodded.

See a pattern? Using tasty treats generously to *create* a positive association with many potentially aversive stimuli and experiences is **classical conditioning**—as opposed to counter conditioning, which is changing an already *existing association*, usually from negative to positive.

Present any potentially aversive stimulus at a low intensity at first— far away, low volume—and associate it with yummy stuff. Every time the scary thing appears, feed your foster dog tasty treats—lots! As long as they can see the scary thing, feed tidbits. When the thing leaves, stop feeding. When your pups realize that the thing *makes* good stuff happen, they'll *want* the scary thing to appear. Then you can increase the intensity—closer, louder—and keep feeding treats, until the pups are completely happy about the sight or sound.

Do this with things that are neutral, making sure your fosters have a positive association with those things too. Children are the most common victims of dog bites. Start creating your pups' positive kid association the first time they see children by feeding treats *whenever* young humans are around—don't wait to find out if they are afraid.

Does fostering puppies sound like a lot of work? It is, if you do it right. Ideally the litter should stay together and with their mom until they are eight weeks old. They need to learn things from mom, and from each other. Pups taken away from their litters early are often too hard with their teeth, because their mother and their littermates help them learn bite inhibition. If they must be separated sooner for some reason, they should at least stay with littermates—three or four to a group. They'll still have the benefit of each other to teach and learn from. As part of your socialization process, occasionally remove each from the others, and let them become accustomed to being alone. This will decrease the stress—and crying—when they eventually go to their forever homes with their new humans.

Note—One legitimate reason for taking pups away from mom early is if she has health issues and is unable to care for some or all of them herself. Another is if she is fearful herself and/or aggressive. Her pups will learn to be fearful and aggressive by watching her, and can be weaned early to avoid this.

10 prime socializing opportunities

After your new foster pup has had a day or two to settle in, you can begin active socialization work. Remember, it is critically important that your pup have *positive* experiences when you are socializing—take good care not to allow her to be frightened or intimidated! If an environment seems to be overwhelming her, abort the session and next time look for a quieter location. Here are ten socialization suggestions:

1. **Visitors.** Invite friends over, one or two at a time, for cookie parties. Have everyone sit on the floor with tasty treats in hand, and let the pup visit. Make sure everyone knows this important rule: the pup gets to decide who to visit, and for how long. No grabbing, restraining, or picking up the puppy. Try to vary the type of visitor—different ages, races, and cultural styles.

2. **Service people.** Hand your mail carrier some treats and ask him to feed your pup. Same with the pizza delivery guy, UPS and Fed Ex delivery folks, and any military personnel who might be around.

3. **Riding in cars.** Cars can be scary places for puppies. They make noise, they move, they cause carsickness, they take pups to scary places like veterinary hospitals… Play car games with your non-running stationary vehicle until your pup thinks the car is the best place in the world. Sit with her in the car and feed her treats, encourage her to climb in (if she can reach) and feed her treats. Let her lie in the back seat next to you, chewing on a food-stuffed toy while you read a book. When she thinks the stationary car is the best thing in the world, turn on the engine and play the same games until the engine noise doesn't bother her. Then drive a short distance—10 to 20 feet, hop out, play her favorite game, hope back in, drive back to your starting place, hop out and play again. Gradually increase the length of the drive, playing games at each stopping place.

4. **Neighborhood scenes.** Starting in a quiet neighborhood, just hang out on the sidewalk with your pup and watch the world go by. Be sure to take along plenty of treats, so you can have people feed her, and so you can feed her if she appears worried about anything.

5. **Town life.** If all goes well in your quiet neighborhood, try the busier world of a small town. More sights and sounds to absorb… remember it has to be positive and fun for her!

6. **Puppy class.** A well-run positive reinforcement-based puppy class is the best place for your foster pup to meet other canines. Your knowledgeable instructor will make sure she gets to play with other appropriate canines so she isn't frightened by an unruly playpartner.

7. **Stores.** Do a little research to find out what stores in your area allow dogs to enter. Avoid pet supply stores until your pup has some socialization miles under her belt at the abovementioned well-run puppy class. Even then, be prepared to protect her from the advances of unruly dogs. Meanwhile, some hardware stores and some locally owned shops welcome dogs.

8. **Wildlife/Livestock.** A friend with a farm where you can hike and meet deer, squirrels, bunnies, cows, sheep and horses is the perfect opportunity to teach your on-leash pup that all those amazing creatures make high-value treats appear in your hand. This will lay a solid foundation for her to understand that other animals are for getting treats, not for chasing.

9. **Water.** Get a kiddie wading pool, put an inch of water (or less!) in it, and feed treats/play games with your pup in the water. Gradually increase the depth of the water until she is happily splashing and swimming. Graduate to a quiet pond, if you have access to one. Stay away from crashing waves and fast-moving rivers!

10. **Public events.** Public events such as carnivals and street fairs are advanced socialization opportunities, Start small, with a neighborhood yard sale or picnic, where there won't be scary bands, costumes, loud-speakers and fireworks. If your pup does well with those, perhaps a grade school soccer game—but don't let the kids mob her!

Remember, when you are socializing your foster pup, err on the side of caution; it only takes one scary event to create a lifetime phobia. Don't be afraid to tell people to back off if your pup seems reluctant to meet them. If you can see she is worried, back away from the scary thing. If she doesn't relax, take her home, and make a mental note to go slower with the socialization. If she seems frightened of many or most things when you are trying to socialize her, let your organization know that you may need the help of a behavior professional.

Record keeping and reporting

A foster dog can require a surprising amount of paperwork, from vaccine records to receipts. It is important to keep good records of all you do for and with your foster dog, starting with medical records. He should come to you with his current vaccinations and any necessary de-worming already administered. Be sure your shelter or rescue agency provides you with documentation of those for your files. Make sure *before* you pick him up that they have indeed had vaccinations taken care of, and acquired a health certificate if he will be crossing state lines. If this has not already been done, arrange to do it on your trip home with him (*before* you cross into the next state) or, if they are delivering him to you, as soon as he arrives. If you live in an area with ticks and/or mosquitoes, also make sure he is tested for and protected against Lyme disease and heartworm or, again, immediately do that yourself. If there are any other potential health threats specific to your area, be sure to check on those as well. Some breeds of dogs—notably Australian Shepherds and Rough Collies, but some others as well—may be sensitive to some heartworm medications. Ask your vet if your foster is one of the high-risk breeds, and have her suggest alternatives.

Fleas can be another hazard when you bring a dog into your home from random sources. If your group doesn't do routine flea control before delivering fosters, do it immediately yourself. Trust me, you do *not* want fleas to infest your home. I've been there, and done that, and it isn't fun.

There is currently some controversy over the use of topical flea control products. Some dogs are sensitive to their use and their reactions can range from mild annoyance (running around, rubbing and rolling) to the development of mild to severe burns at the application site. Have discussions with your partner organization and your veterinarian to see what they recommend for effective and safe flea control.

Make sure you keep all receipts for any expenditures. If your agency is reimbursing you for costs, they will need copies. If they are not, check with your accountant—it's likely that your unreimbursed expenses incurred during your volunteer work for a 501(c)3 non-profit group are tax-deductible.

I urge all canine foster parents to also keep a behavior journal, preferably in a word processing program that you can print out. If you are willing to keep one, it will be a valuable gift to your foster's new owners when he eventually is adopted. Reading your past entries will also assist you in evaluating how quickly and well he is adapting to his foster situation, and help you decide if you need to do any behavioral interventions while he is with you. Documenting all this will make his adjustment to the next change in his life easier. His new humans will have a better idea of what to expect, and the record of your experiences will help them know how to work with any challenges that may arise. It is critically important that you are forthright about any issues you encounter with any and all of your foster dogs. Hiding difficult behaviors, while it may get dogs adopted more easily, will also make it more like that they'll be returned. You want fully informed adopters who want and love their adoptees, behavioral warts and all.

Sample journal entries

Your early entries in your journal might look something like this:

Day 1. Mitzi approached the car easily, without any signs of fear, and jumped in of her own volition. It took a few treats to coax her into the crate. After a few minutes of whining in her crate she lay down and rode home comfortably for the remainder of the 45-minute car ride from Anytown Humane Society to our house. We introduced her to our dog, Rufus, in the fenced back yard without incident—they played happily together for about a half hour. Mitzi likes to be the chaser more than the chasee, and Rufus was happy to oblige. We then brought them in and fed them dinner (in separate rooms, to avoid any trouble) and settled in the living room for the evening. We took her out every hour to eliminate—don't want any accidents, and don't know yet if she is housetrained. It again took a few treats to coax her into her crate in our bedroom for the night.

Day 2. Mitzi slept almost through the night—waking us up just before 6:00am whining. I took her out and she immediately went to the bathroom. I brought her back in and re-crated her until our regular 7:00am wake-up time. Nothing remarkable today—we reduced potty breaks to every two hours, still with no accidents, so perhaps she is housetrained—although other than the whining in the crate she's not giving any signals to go out. I always say "Let's go outside!" when I take her out. She shows friendly, appropriate interest in our two cats, so would probably be a good candidate for adoption in a cat-owning home.

Day 3. Daughter came over today with four-year-old grandson. Rufus loves Jordan, but Mitzi, not so much.

She moved to the far side of the room and kept at least 15 feet away from him at all times. We didn't let Jordan try to approach her, and will recommend to the shelter that she needs to be placed in a home without small children. In the future when Jordan comes over I will probably put Mitzi in her crate in our bedroom and close the door, so she doesn't have to worry about him.

Day 4. Yay—today Mitzi told me she had to go outside! I was sitting at my computer and she came and woofed at me. When I said "Do you have to go out?" she barked and ran to the back door. I took her out and she immediately peed. Good girl! We are now feeding her and Rufus in the same room—opposite sides of the kitchen, without incident—although I do monitor them to make sure whoever finishes first doesn't harass the other.

Since Mitzi doesn't appear to have had any training, I signed her up for a basic good manners class with a positive reinforcement trainer. Class starts next Wednesday evening.

Day 5: et cetera...

You will want to report to your shelter or rescue group on a regular basis. Hopefully they have a person designated as the foster coordinator, or at least have a specific person you report to, so there is consistency in your communications. The sooner you let them know that Mitzi is good with cats, is housetrained, and doesn't like small children, the better they can focus their efforts toward finding her perfect forever home. If other behavioral issue arise, especially serious ones, you need to tell them immediately so they can assist you in getting appropriate behavior modification help.

When it doesn't work

Despite your best efforts and intentions, there will be times when it just doesn't work out for the foster dog you've welcomed into your home. Often the handwriting is on the wall in just a couple of weeks. Perhaps you babysit for your grandson Jordan every day when your daughter goes to work, and there's simply too much risk in trying to keep Mitzi and Jordan under the same roof 40-plus hours a week.

You would be doing Mitzi a huge disservice to try to manage the situation. Management has a high risk of failure, especially when there's a toddler around who can open doors and wants to play with the nice doggie. One bite to a child, and Mitzi's prospects for adoption plummet, while her chances for euthanasia skyrocket. And that's not even considering the potential for serious injury to Jordan! It's just not worth the risk.

There are a multitude of other reasons why your current foster might need to be returned. Perhaps he barks incessantly while you are at work all day, and you have nearby neighbors who are complaining. Perhaps he has isolation or separation issues, and needs to be with a more behaviorally experienced foster parent (who doesn't work all day) who is better able to work with this. Maybe he and Rufus *don't* get along (or he is dedicated to chasing your cats) and you realize it's too stressful to your other four-legged family members to keep him around. Whatever the reason, returning him sooner, rather than later, makes it more likely that your group will be able to find a different foster situation that is better suited to his needs. You will be sad, but you will also then be able to take on a different foster who will be able to thrive in the loving environment you can provide.

4

Care, Behavior and Training
of Your Foster Dog

You will probably want to give your foster dog the best of everything. Food, vet care, training, attention…after all, why not? You are committed to doing everything you can—so why not give him give him the care and training he needs to have the best chance possible for a long and happy life?

Caring for your foster dog's physical needs involves a number of things including:

- Food/diet
- Grooming
- Veterinary care
- Equipment and training tools
- Exercise

Food/diet

Your idea of "the best" dog food may differ from that of the shelter or rescue you are working with. Perhaps you believe "raw" is the best of the best, and the only way to go. Your agency may not want you to feed raw. They may have access to donated food, or a good deal on purchased food because they get it in large quantities. They might want you to use their food, so they don't have to pay extra for your dog's meals. Even if you are willing to pay for it yourself, they may want to keep him on their food so his diet is consistent. If you want to stay on good terms with your foster organization, you'll need to abide by their wishes—unless you can educate them about the importance of good nutrition for dogs and get them to agree with your feeding choices.

If they do agree to give you free rein on feeding, be sure your foster dog comes with a week's supply of the food he's been eating, and then switch him over gradually to your diet of choice. Sudden dietary changes can cause gastrointestinal upset—read: *diarrhea*—and that's the last thing you need to be dealing with as you help your foster settle into his temporary quarters. When it's time for him to go back to the shelter—or to his new home—send a week's supply of *your* food along with him, so his new caretakers can switch him gradually to his *new* diet plan.

While a top-quality diet is certainly the best option for any dog, plenty of dogs survive in perfect health on low- to medium-grade foods (the kind available at most grocery stores)—so it's not the end of the world if your group insists you feed their less-than-ideal diet. If you are allowed to choose, here are some things to look for, to help you separate great foods from less-great foods:

- **Lots of animal protein at the top of the ingredients list.** Ingredients are listed by weight, so you want to see a lot of top quality animal protein at the top of the list; the first ingredient should be a "named" animal protein source (see next bullet).

- **A named animal protein—chicken, beef, lamb, and so on.** Generic "Meat" is an example of a low-quality protein source of dubious origin. The label should say "Chicken," "Beef," "Lamb," etc. Animal protein "meals" should also be from named species; look for "chicken meal" but avoid "meat meal" or "poultry meal." Also avoid "by-product" in relation to protein source—no "chicken by-product," "lamb by-product," etc.

- **When a fresh meat is first on the ingredient list, there should be an animal protein meal in a supporting role** to augment the total animal protein in the diet. Fresh (or frozen) meat contains a lot of water, and water is heavy, so if a fresh meat is first on the list, another source of animal protein should be listed in the top three or so ingredients.

- **Whole vegetables, fruits, and grains.** Fresh, unprocessed food ingredients contain nutrients in all their natural, complex glory, with their fragile vitamins, enzymes and antioxidants intact. Don't be alarmed by one or two food "fractions" (a by-product or part of an ingredient, like tomato pomace or rice bran), especially if they are low on the ingredients list. But it's less than ideal if there are several fractions present in the food, and/or they appear high on the ingredients list.

- **A "best by" date that's at least six months away.** A best by date that's ten or eleven months away is ideal; it means the food was made very recently. Note: Foods made with synthetic preservatives (BHA, BHT, ethoxyquin) may have a "best by" date that is as much as two years past the date of manufacture.

The following are things you *don't* want to see in the ingredients.

- **Meat by-products, poultry by-products or feather meal.** Higher-value ingredients are processed and stored more carefully (kept clean and cold) than lower-cost ingredients (such as by-products) by meat processors.

- **A "generic" fat source such as "animal fat."** This can literally be any fat of animal origin, including used restaurant grease. "Poultry" fat is not quite as suspect as "animal fat," but "chicken fat" or "duck fat" is better (and traceable).

- **Added sweeteners.** Dogs, like humans, enjoy the taste of sweet foods. Sweeteners effectively persuade many dogs to eat foods composed mainly of grain fragments (and containing little healthy animal protein).

- **Artificial colors, flavors, or preservatives (i.e., BHA, BHT, ethoxyquin).** The color of the food doesn't matter to your dog. And it should be flavored well enough to be enticing with healthy meats and fats. Natural preservatives, such as tocopherols (vitamin E), vitamin C, and rosemary extract, can be used instead. Note that natural preservatives do not preserve foods as long as artificial preservatives, so owners should always check the "best by" date on the label and look for relatively fresh products.

- **Ingredients from China.** Some of the largest and worst contaminated food cases have involved products that contain foods from China. Avoid all foods that contain products from China. The country of ingredient origin is often *not* listed on the bag or can, so you may need to do some research. Check websites or contact the company live to confirm their products are China-feed-free.

INGREDIENTS

Deboned Chicken, Chicken Meal, Turkey Meal, Brown Rice, Peas, Barley, Sweet Potato, Chicken Fat (preserved with natural mixed tocopherols), Salmon Meal (source of Omega 3 fatty acids), Oats, Natural Chicken Flavor, Carrots, Apples, Blueberries, Organic Alfalfa, Minerals (Salt, Dicalcium Phosphate, Calcium Carbonate, Zinc Amino Acid Complex, Zinc Sulfate, Iron Amino Acid Complex, Manganese Amino Acid Complex, Copper Amino Acid Complex, Potassium Iodide, Cobalt Amino Acid Complex, Sodium Selenite), Vitamins (Choline Chloride, Vitamin E Supplement, Vitamin A Supplement, Vitamin B12 Supplement, d-Calcium Pantothenate, Vitamin D3, Niacin, Riboflavin Supplement, Biotin, Pyridoxine Hydrochloride, Folic Acid, Thiamine Mononitrate), Yucca Schidigera Extract, Dried Lactobacillus plantarum fermentation product, Dried Lactobacillus casei fermentation product, Dried Enterococcus faecium fermentation product, Dried Lactobacillus acidophilus fermentation product, Rosemary Extract.

Check the ingredients label to be sure your foster is getting good quality nutrition. This is the label from a good-quality food.

Feeding schedule

Most dogs need to eliminate shortly after eating, so be sure to schedule your dogs' meals at a time when you can give them a potty break 20 to 30 minutes later (sooner for puppies!). You want to be sure they are "empty" when you leave for work, and when you go to bed at night. For this reason, and for several others, I am a major fan of feeding dogs regular meals, rather than free-feeding:

- If you have multiple dogs in the household and you free-feed (fill a bowl with food and put it on the floor, refilling as needed) you can't know, or regulate, how much each dog eats.

- Free-feeding risks resource guarding issues, with possible fights, and less assertive dogs not getting enough nutrition (and being very stressed) because they are being kept away from the food by more assertive dogs.

- You don't know when a dog is "off his feed" if food is available all the time. My dogs all lick their bowls clean within ten minutes, maximum, of bowl delivery (usually five minutes). If a dog doesn't empty his bowl one meal I'm on high- alert for a health issue. If two meals are missed, I'm on the phone to the vet.

- Housetraining goes more smoothly with meals. It's easier to manage output if you control input.

- You can control weight more easily with meals. Thin dogs get more food, overweight dogs get less.

- You can use mealtimes to training advantage, either by using the dog's meals for training sessions throughout the day, or by scheduling meals to *follow* training sessions, taking advantage of your dog's empty stomach to maximize the impact of your positive reinforcement training.

Grooming

Your foster dog may—or may not—tolerate necessary grooming procedures. This is a great time to help him learn that activities such as grooming, nail-trimming and tooth brushing are enjoyable, not scary. If your dog has been coerced into submitting to these procedures in the past he may become quite violent when you attempt to brush his coat or teeth, or trim his nails. The key is to give him a positive association with the tools involved, and only very slowly actually begin using them.

Start by simply determining if your foster is sensitive to being touched anywhere. Gently touch and pet him over all the "normal" places. Scratch under his chin, stroke his back and sides, rest your hand on top of his head, look in his ears, lift his lips to look at his teeth, and gently lift each paw. Jot down a note about any places where he becomes tense, pulls away, or worse, growls or snaps. Do *not* punish him for growling or snapping. He is trying to tell you, as politely as he knows how, that he doesn't like what you are doing. Rather than punishing him for communicating, respect his message, and commit to helping him become more comfortable with being touched. Punishment will only confirm his conviction that being touched is bad. Counter conditioning and desensitization (CC&D), on the other hand, will convince him that being touched is *wonderful*. You will find more details on CC&D program in the section that follows.

It's almost magical to watch an effective CC&D program in progress. Some behavior changes I've seen as a result of this kind of behavior modification have been nothing short of miraculous, such as one family's Chow mix and newly adopted Chow who wanted to tear each other to shreds, but became fast friends within three weeks when the owners implemented a CC&D program.

Reprogramming for grooming tolerance

Perhaps your foster dog will tolerate a light touch on the top of his head and a gentle scratch under his chin, but if he becomes very tense if you do more than that and if efforts to touch his legs and feet or his hindquarters and tail elicit serious warnings, be on the lookout for forthcoming aggression. You must believe him and wisely don't press the issue. View even simple but necessary procedures such as nail trimmings and baths as stressful and dangerous. Your dog needs some reprogramming!

The first step in your program is to have a complete and thorough veterinary exam and an adjunct visit to a chiropractor if your veterinarian isn't skilled in that practice. Pain is a huge contributor to aggression—if he's hurting, all the CC&D in the world won't change his opinion of being touched—*it hurts!*

Session 1. With a clean bill of health, you're ready to begin. You'll need a large supply of absolutely scrumptious treats—canned chicken, rinsed and drained, is my favorite for CC&D purposes—most dogs totally love it. Pick a comfortable spot on a bed that your dog loves, or lay down a cushion or a soft thick blanket for the two of you to sit on. Attach a leash to your dog's collar so you don't have to grab at him to keep him with you.

The sequence of the next part is very important. You will touch your dog's head *first*, very briefly—say for one second—and feed him a tiny bit of chicken while your hand is still touching him. The touch must come first because you want him to understand that *the touch makes the chicken happen*. If you feed chicken first, then touch, he won't make that connection. Keep repeating this step until your touch causes him to look at you with a smiling face as if he's saying, "Allright—you touched me. Yay!!! Where's my chicken?" You want the "Where's my chicken?" (WMC) response to happen reliably several times in a row before you proceed to the next step.

Good job—you've accomplished the first tiny step on a long road—he thinks being touched softly and briefly on the head is a wonderful thing! Now you must decide whether to stop the session—ending on a high note—or continue on because you both are having a wonderful time and don't want the session to end. If you're unsure how much longer he will work with you, it's better to stop sooner, while you're ahead, than to push it too far and suffer a setback.

If you proceed, the next step might be to touch him on the head, still very gently, but for two seconds. You may lose the WMC response at first as he adjusts to the increased time, but it will probably return quickly. Continue to increase the time very gradually so you don't lose the progress you've made. As the touches get longer, feed him several treats in rapid succession *while* you are touching. Remember to stop the treats when the touch stops.

Be sure to end the session before one or both of you gets bored, tired, stressed or frustrated. You can always do another session later today or tomorrow. If you sense that he's thinking about getting restless, stop the session, feed him a few extra tidbits for being a wonderful boy, and release him with an "All done!" cue. Next time, stop a little sooner—you don't even want him to *think* about getting restless.

Session 2—Taking the next step. When you start up again with your next session, back up a little. If you ended with five-second gentle touches on your dog's head, start with two-second touches. You'll be able to progress more quickly back up to five seconds, but you want to be sure to start within his comfort level and warm up to the place where you ended.

When he has a positive association with gentle touching up to perhaps ten seconds, you can increase the intensity of a different stimulus—the amount of pressure. Each time you raise the bar for a new stimulus, lower it for the others—in this case you might go back to two or three seconds, with a slightly stronger pressure when you touch. Work to get that positive WMC response with the new amount of pressure at each length of time before you increase the time again.

When he's responding happily to a moderate amount of touch pressure at ten to fifteen seconds, you can increase the intensity of the third stimulus in the touch package—the position of your hand. Up until now you've been touching him in his most accepting spot—the top of his head. Now you're going to begin to move your hand to more sensitive places—again reducing the intensity of the other two stimuli—time and pressure.

Perhaps you'll try ears first. Returning to a very gentle touch, stroke one ear for one to two seconds, and feed chicken while you stroke. Repeat this until you're getting his WMC response to the ear-stroking, then do the same with the other ear. Gradually increase the length of time you stroke each ear gently, and when you're getting positive responses to ten-second ear stroking, it's time to increase the pressure. Shorten your ear strokes back to one to three seconds, but stroke the ear a bit more firmly. Remember to be very generous with your chicken bits—feeding a morsel or two every time you stroke the ear, and several morsels as the touches get longer. When he's happy to have you stroke both ears firmly for ten to fifteen seconds or longer, you can move to a new spot.

Don't forget to reduce the other stimuli each time you move to a new touching place. After the ears, you might run your hand down the back of his neck, gently and briefly. Treat as you touch! You should find that as you work toward various new spots around your dog's legs and body, he'll accept new touches more quickly in many places. Adjust your pace to his behavior. If he's giving you WMC responses very quickly, you can progress more rapidly in your program. If he seems slower to respond, you're probably working on or near a very sensitive place, and you need to slow the program down. He'll tell you how slowly or quickly you can progress. Listen to him. Attempts to force him to accept your touching will backfire, bigtime.

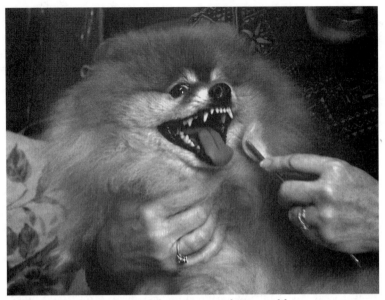

This Pomeranian objects to being groomed. He could use some counter conditioning and desensitization work. Photo: Shirley Greenlief

Sensitive places

Many dogs, even those who are comfortable being touched elsewhere, are tense about having their feet handled. Take extra care as you begin to move down your foster dog's legs. A few extra days—or weeks—now will pay you jackpots in the long run, when you can finally clip his nails without a violent struggle. Spend lots of time massaging the areas where your dog has come to enjoy being touched, and occasionally work on the more sensitive spots. In addition to the chicken, soothing massage adds a powerful positive association to the message that you are planting in his brain.

Remember that it's critically important to avoid triggering the negative associations outside your CC&D sessions. If you forget about his sensitivities and grab him during a "real-life" moment you can set your program back. It won't hurt to skip one or two nail-trimming sessions while you work to get him to accept foot-handling without a fight.

*Extra care invested now to help your foster enjoy paw-touching
will pay off big dividends in the long run.*

Make sure that others are aware of the importance of respecting the
CC&D program too. There's nothing like having a friend or family
member think it's funny to see your dog's negative reaction when
they play "grab your paws"—thereby undoing all the good work
you've done. Grrrrr! I've been known to banish human acquaint-
ances from my household for less!

In his own time
How quickly you complete your grooming and touching CC&D program depends on several factors:

- Your foster dog's age and how long the behavior has been happening.
- The intensity of his negative association with touch.
- The cause of his sensitivity. Prior harsh handling is likely to be more difficult to overcome than lack of handling since he has a negative association with the human presence as well as the sensitivity to touch itself.
- Status of physical contributors to the sensitivity—if your dog has a grass allergy that causes inflammation in his pads, for example, your constant struggle to reduce the discomfort in his feet will slow your CC&D progress.
- Your commitment to implementing the program on a daily basis. Several short sessions a day are generally more effective than one long daily session.
- Your skill at reading your dog's comfort level and moving the program forward at an appropriate pace without triggering negative reactions.

Success!
The success rate for touch CC&D programs is high. Unlike modification programs for things like dog reactivity, where it's difficult to control all the variables, you can manage the factors of a touch modification program with relative ease. Chances are good that even if you don't achieve 100 percent positive association with touching every part of your dog's body, you can accomplish a positive response for much of it, with agreeable acceptance for the highly sensitive parts.

Veterinary Care
This section will be short, as I am not a veterinarian and it would be inappropriate for me to try to give specific advice about what veterinary care is needed or appropriate for your foster dog. Here are some general suggestions:

- Rabies vaccinations are required by law in most places (if not all) in the US. Your foster dog *should* have this shot *before* you take custody of him (unless he is still too young a pup for shots). If it ends up being a long-term foster (over a year), you will need to be sure he gets his first rabies booster twelve months later. After the first-year booster, additional boosters are required on a one-, two- or three-year basis, depending on where you live; laws vary from state-to-state.

- The need for other types of vaccinations varies, depending on your location, your dog's age and prior medical history. Discuss this with your veterinarian and foster agency. There is an increasing trend across the country toward less frequent and fewer vaccinations; you may want to do some online searches for resources that discuss this subject.

- Deworming can, again, be region specific for some internal parasites, although roundworms and tapeworms, at least, are pretty universal. Do a little research to find out what other parasites might be common in your region.

  - Puppies are most likely to be susceptible to roundworm infestation (they get them from their mothers) and are pretty much routinely dewormed. Check with your organization to see if your foster pups have been checked and treated for roundworms. You may need to provide a stool sample to your veterinarian. Signs of roundworms include a bulgy "potbellied" appearance, general failure to thrive, a poor coat and spaghetti-like worms in stool or vomit.

  - Tapeworms come from fleas that are ingested by the dog—either by chewing at and swallowing fleas that are on his own skin, or by consuming fleas on captured prey animals such as bunnies and squirrels. (Yes, dogs are dogs, and many of them, given the opportunity, do chase, catch, kill and eat small prey animals.) Tapeworms also cause an unthrifty appearance (although not usually the potbelly look) and you may see dried segments in feces or on the dog's anal area that look like small white grains of rice. Live tapeworms are flat, segmented, white, with a triangular head. Again, you may need to provide a stool sample, although if you report and describe seeing tapeworms, most vets will dispense dewormer without requiring a fecal sample.

- It is appropriate to use flea and tick control measures in many parts of the world, both for the dog's own comfort as well as to protect from tick-borne diseases such as Lyme, Rocky Mountain spotted fever, and more. Have a conversation with your veterinarian and foster organization to determine what measures are needed in your part of the world, and which ones are appropriate for your specific foster dog. Some breeds are sensitive to some products.

- Blood tests are necessary to confirm the absence (hopefully) or presence of heartworm and Lyme disease, among others, prior to treatment or the use of preventatives. A senior blood panel is recommended for older dogs, and may also be necessary for younger dogs to identify an undiagnosed illness. Communicate with your group and your veterinarians if you think a blood panel is indicated or recommended, and follow their instructions. I suggest having a good medical resource in your library, such as *Merck's Veterinary Manual*.

- Minor medical needs can happen anytime. A cut of unknown origin on the paw pad. Itching. Ear mites. Dental needs. Have an understanding with your group, in writing, that specifies how minor medical needs are handled.

- Pregnant females require extra care and attention. If you are thinking of caring for a mom-dog-to-be, talk with your vet and the rescue group about what to expect, what to look for, and what additional care will be needed.

- Protocol for handling catastrophic medical needs also must be addressed before they occur. Hopefully they won't, ever, but despite your careful attention, life-threatening and debilitating emergencies can and do happen. Torsion (bloat). Broken limbs. Hemorrhagic gastroenteritis. Cruciate ligament tears. Ingestion of toxins. The list, unfortunately, is endless. In the *unlikely* chance that something serious does happen, a prior understanding, *in writing*, of how to proceed can take at least *some* of the angst away from a very angst-generating experience. You need to know:

  - Will your foster organization pay for emergency care?

  - If so, is there a cap on spending for an emergency event?

  - Must you go to the group's designated veterinarian, or can you go to your choice in an emergency?

- If you must use their designated vet, what if it's after hours, or weekends, when their vet isn't available? What if their vet is too far away?

- Who do you contact to make emergency treatment decisions? What if that person cannot be reached? Can you then make decisions, or do you need to contact a back-up person?

- If the cost will exceed the group's spending cap, can you pay for treatment yourself?

- If treatment is failing or appears futile, who can make decisions about euthanasia?

Equipment and training tools

Tags and a flat collar

You can never have too much identification on your foster dog. Well, maybe that's a bit of an exaggeration, but you certainly want to be able to get him back as easily and quickly as possible, should he manage to slip away from you. At minimum, a tag with your contact information and one with the contact information for your organization, and a license, if one is required, should be on his flat buckle collar *at all times*—and he should be wearing that collar *at all times*. If you are concerned about the small possibility that tags could get caught in crate wires or heating vents, have them riveted to his collar so they don't dangle.

Microchips

Permanent identification is also useful. Many shelters and rescues now routinely microchip their canine charges, and most shelters around the US have scanners with which they check all incoming animals. If your foster is not already "chipped" when he comes to you, ask your group about the possibility of having it done for extra safekeeping.

Tattoos

Tattooing has somewhat fallen out of favor for identification purposes, because tracking tattoos often proves difficult. If your foster *is* tattooed, find out if your group had it done. If not, ask if they made an effort to trace the tattoo back to an owner. If they didn't, you have a detective job to do. Google "dog tattoo" and get started! If they did try to trace it, they may have been unsuccessful.

If your group had the tattoo done themselves, ask where the tattoo is registered, and note the tattoo and registry contact information in your foster records. If he does go missing, you'll want to have access to it so you can let the registry know he is missing, and to contact you if he is found.

GPS

Yes, you can now purchase a GPS tracker for your dog. I have one called Tagg–the Pet Tracker. This is a rechargeable battery-powered unit that attaches to a special collar that comes with the tag. You register the dog on your computer, and if he escapes you can track him on a computer or smart phone (there's an app for that). Speaking from personal experience, I can attest to the peace of mind this product provides if you have a dog prone to going walkabout.

Since you will probably have a succession of foster dogs over time, you could register your Tagg to "Foster." When your current dog finds his forever home, just remove the collar and Tagg and put it on the next one!

The Tagg Pet Tracker—a GPS for your foster dog…not a bad idea!

Exercise

One of the easiest and most effective things you can do to benefit your foster dog is to give him plenty of exercise. A surprising number of canine behavioral challenges can be mitigated simply by the addition of a good exercise program. Let me repeat that. A surprising number of behavior challenges can be mitigated simply by the addition of a good exercise program. Many of our fosters who came with some aggressive behaviors related to low tolerance for frustration and low impulse control, fared as well as they did because we could take them for hikes several times a day on our 80-acre farm.

Off-leash hikes are the best exercise option for
most dogs if you can do so safely and legally.

Exercise does not just use up energy your foster might otherwise apply to unwanted behavioral endeavors; a good aerobic exercise session causes a release of mood-enhancing, feel-good endorphins, just like it does for humans. It makes him happier—and that's a good thing!

If you don't have access to a safe off-leash run-until-you-drop hiking area for your foster, look for other alternatives for real aerobic exercise. You might try using a **long-line**—a 15- to 50-foot leash that gives your dog more freedom to run. A walk on leash is not sufficient exercise for most dogs. In fact, a walk on leash is an exercise hors d'oeuvre for most dogs, and can be very frustrating for them. I suggest to many of my clients that they exercise their dog in the back yard with twenty minutes of Frisbee or ball-chasing before they go for a walk on leash, and I'll suggest the same to you. You and your foster dog will enjoy the walk more if the dog isn't exploding with energy.

If you don't have access to a safely enclosed, legal off-leash area, try exercising your foster on a long-line to give him more room to run.

Mental exercise can be as tiring as physical exercise for many dogs. Invest in some interactive puzzle toys, get a good book on shaping, play nose games with your dog, and you may be surprised by how effectively those activities take the wind out of his sails.

Six exercise caveats

While exercise is, indeed, a cure-all for many behavioral ills, there are times when it's *not* the right thing to do, and there are some kinds of exercise that are better avoided. Here are some cautionary notes to keep in mind as you plan your foster dog's exercise program:

1. **Start slowly, and check with your dog's veterinarian before you start.** Just as with humans, a dog who is unfit or overweight can be physically harmed by too much exercise. These dogs need to build gradually to more strenuous exercise sessions.

2. **Be aware of any limiting physical conditions.** If your foster had recent surgery, exercise needs to wait until he is healed. Follow the veterinarian's instructions. If your dog shows any signs of lameness or stiffness as you introduce his exercise program, a visit to the vet is in order to determine if it's just muscle soreness, or if he has a more serious medical condition such as hip dysplasia, arthritis, a torn ACL, etc. If your foster seems reluctant to exercise, have him checked for medical problems.

3. **Watch the heat.** If you are in a hot summer season or just experiencing unseasonably warm weather, exercise your foster dog in the early morning or after dark, if and when things cool off. Dogs can overheat quickly and easily in warm weather, and heat stroke is nothing to play with.

4. **Beware hard surfaces such as asphalt.** Pavement can become extremely hot in warm weather and burn your dog's pads. Also, high-impact games such as Frisbee-catching should be done on a surface that has some give, to reduce the chance of injury and the development of arthritis.

5. **Be careful with Brachycephalic breeds.** These are dogs like Pugs and English Bulldogs who have very flat faces. They have inherent breathing difficulties, and cannot exercise as vigorously as longer-nosed dogs. Ask your veterinarian how much exercise is appropriate for your short-nosed foster dog.

6. **Avoid inappropriate activities that encourage undesirable behaviors.** Games that encourage your foster dog to put his mouth on or jump roughly at humans reinforce behaviors that may get him in

future trouble. Play games and exercise with him in ways that reinforce desirable behaviors. Playing tug with a designated tug toy is great. Playing tug with clothing or human body parts is not. Jumping over jumps is terrific. Jumping roughly on humans is not.

If you have any doubt about whether a particular activity is appropriate or not, ask your friendly neighborhood positive reinforcement trainer, or contact your foster organization and ask them.

Use caution when exercising brachycephalic (flat-faced) dogs.

Behavior and basic training

While you may only have your foster dog in your home for a relatively short time, knowing the basics of dog behavior and how to train simple behaviors will help ensure the time the dog is with you goes as well as it can. Plus it provides a real leg up for whoever adopts the dog on a permanent basis. By being proactive and providing the dog with a solid training foundation you can avoid many of the problem behaviors that, unfortunately, may arise. We will cover how to deal with problem behaviors in Chapters 5 through 9.

Dog behavior 101

If you are going to be working with dogs, especially dogs who may be in need of some rehabilitation, it's important that you have a working knowledge of some basic behavior principles. This will help you avoid the pitfalls of some of the off-the-cuff advice you may get from friends and family, as well as some of the dangerous misinformation you can find on television—information that can actually do more harm than good.

First, and most important, *it's not about dominance.* (See Chapter 5—Problem Behavior: The Myth of Dominance and Alpha.) You don't need to show your foster dog who's boss. Physically coercing a dog to do things, holding him down, hitting him with your foot, staring him in the face…anything incorporating forceful or pain-causing handling, training methods or tools—such as choke chains, prong collars or shock (electronic) collars—can easily cause intimidation or fear in the dog; this is the last thing *any* dog needs, but particularly a foster dog whose world is already out of control, turned upside-down and full of stress. It can also result in an aggressive response from the dog—and dogs who bite people tend to have short lives.

Dog behavior, as well as that of many other species, including humans, combines these four concepts:

Reflexive behavior. Reflexive behavior is unlearned or unconditioned behavior. When you touch your finger to a hot stove burner, you yank it back because it hurts. It's a reflex. No one had to teach you that. When you step on your dog's paw, he yelps and moves away because it hurts. No one had to teach him that.

Classical (associative/respondent) behavior. Classical conditioning is about making associations between stimuli. When you accidentally step on your dog's paw—or worse, deliberately use physical punishment—he learns to associate your close proximity to him with pain. Socializing your young foster pup will provide a *positive* classical association with a wide variety of things in his world so he grows up with an optimistic attitude about the world. Implementing a classical counter conditioning program with a fearful or undersocialized adult dog can give him a new, more positive association with things that frighten him. *Negative* classical associations can lead to aggression. Dogs who are trained with forceful, fear or pain-causing methods can develop a negative classical association with training, and sometimes fight back. If a dog has bad experiences with young children, he can learn to growl at and bite children in order to keep himself safe. Classical associations and the emotional responses that accompany them are not under the dog's deliberate control, any more than your own classical associations are. Are you terrified of spiders or snakes because some cruel playground bully dropped one down the front of your shirt at recess? Try to control your response when one crawls across your foot. *That's* classical conditioning.

Operant behavior/conditioning. With operant conditioning, a dog makes deliberate behavior choices in anticipation of the consequences of that behavior. A dog's goal in life (similar to ours!) is to make good stuff happen and bad stuff go away. He chooses to sit because he knows sitting often makes treats happen. His positive association with treats prompts him to sit deliberately, in hopes of creating the consequence of getting a treat. He avoids being under your feet so he doesn't get hurt. His negative association with the pain of being stepped on prompts him to deliberately stay away from your feet in order to avoid the consequence of getting hurt.

You can use operant conditioning to teach your foster dog good manners while he is with you, and increase his chance for success in his new home.
Photo: Shirley Greenlief

The four principles of operant conditioning

Behavioral scientist B.F. Skinner identified the "quadrants" of operant conditioning, that define the principles of an animal's (or human's) deliberate behavior choices. It's helpful for you to know these, in order to understand what is motivating your foster dog's behaviors. First, a couple of definitions:

Positive: Something is added

Negative: Something is taken away

Reinforcement: Future behavior increases

Punishment: Future behavior decreases

When you combine those terms you get the four principles/quadrants of operant conditioning:

Positive reinforcement. The dog's behavior makes a *good* thing happen; future behavior *increases*, because we all want to make good stuff happen.

Example: Your dog sits, you give him a cookie. He sits more often because he figures out that sits make cookies happen.

You want positive reinforcement to be the primary principle you use in your interactions with your foster dog, because it helps build confidence, as well as better relationships between dogs and humans.

Positive punishment. The dog's behavior makes a *bad* thing happen; future behavior *decreases* (because no one wants bad things to happen).

Example: Your dog jumps up on you, and you knee him in the chest to show him that jumping up causes pain. Your dog jumps up less in the future, and he may also be unwilling to approach you because he learned that *humans* cause pain. (This is not recommended.)

You want to avoid the use of positive punishment because it creates stress for your foster dog, which can lead to behavior problems (including aggression), and it can easily damage his trust in humans. Chances are his trust in humans has already been damaged. You don't want to add to that.

Negative reinforcement. The dog's behavior makes a *bad* thing go away; future behavior *increases* because we all like bad things to go away and stay away.

Example: Your mail carrier pushes your mail through a slot in your front door. Your foster dog barks at the intruder, and the intruder goes away. The dog learns to bark sooner and more ferociously when he hears the mail carrier coming up the walkway, and louder and longer as the mail carrier leaves, because he is convinced his barking is making the intruder leave.

There is some limited use for negative reinforcement in a good behavior modification program, but it must be done carefully, under professional supervision. The "bad thing" should not be presented at a level that causes the dog undue stress.

Negative punishment. The dog's behavior makes a *good* thing go away; future behavior *decreases*, because we all want good stuff to stick around.

Example: Your foster dog jumps up on you and you say "Oops!" and turn your back on him to show him that jumping up makes your attention go away. He jumps up less in the future because he doesn't want you to go away— especially if you remember to give him attention (positive reinforcement) when he sits, or stands quietly with four paws on the floor.

You can use negative punishment with your foster dog to let him know when he's doing an unwanted behavior, but you want to use it far less than you use positive reinforcement. If you find yourself using it a lot, you may need to rethink your management and training program.

Cognition. Then there's the whole field of canine cognition—our dogs' ability to think, process thought, understand concepts and have an awareness of the thought processes of *other* beings (known as *metacognition*).

There was a time, many, many years ago, when scientists and philosophers told us that dogs—and other animals—didn't feel pain. We know, now, how cruelly wrong that was. Then we were told that only *humans* could make and use tools. Now we know that many species of animals, including other primates, birds and others, can make and use tools. Go to YouTube and search "crow, tool" to see some incredible clips of a European crow making and using tools.

Even birds can make and use tools. Don't underestimate your dog's intelligence and cognitive potential.

Then scientists conceded that other animals could make and use tools but, of course, they didn't have human emotions. In fact, how species-centric it is for us to even call them *human* emotions! We know now that they are simply *emotions*, and that dogs and many other species of animals share a range of emotions very similar to ours. Ask any dog owner if her dog can be happy, sad, angry, fearful, loving—and you will get a resounding "Yes!" response. Those are emotions.

That leaves us with cognition. "A dog's cognitive ability is very limited," behavioral scientists told us in the not-too-distant past. "They are long-domesticated, and domestication dulls intelligence."

Once again, scientists are proving themselves wrong. Only in the last ten to fifteen years have dogs even been considered legitimate behavioral study subjects. Now that there are canine cognition laboratories around the world, science is making amazing discoveries about our dogs' ability to reason, problem solve, grasp concepts and just plain think. Rather than dulling intelligence, it now seems more likely that their close association with humans has *enhanced* their mental abilities.

Dr. Brian Hare operates one such canine cognition lab at Duke University in North Carolina. In his recently published book *The Genius of Dogs*, Hare calls it "a revolution in the study of canine intelligence," and gives many detailed examples of our dogs' incredible cognitive abilities.

So don't sell your foster dogs short. As you go through the fulfilling experience of giving your foster dogs a second chance at a first-class life, take the time to give them the best chance possible by working with their incredible capacity to think and learn, using modern, effective, scientific, positive reinforcement-based training that creates relationships between dogs and humans based on mutual trust, respect and love.

Basic training

You can greatly enhance the probability of your foster dog's future success if you provide some basic training while he is with you. Behaviors every foster dog should know include:

- Sit
- Down
- Wait
- Walk on leash
- Polite greeting

Take advantage of one of more of the excellent dog training books listed in the back of this book for detailed information on how to train these and other useful behaviors. You can also look for positive reinforcement-based trainers in your area and sign your foster up for group classes and/or private training.

In addition, it will be very useful if you can crate-train your fosters while they are with you. With all the caveats about over-crating in mind, teaching your fosters to love their crates can be invaluable for a number of reasons:

1. Your foster dog can take his crate with him when he is adopted, which will make the transition much easier; his own familiar "bedroom" magically appears in his new home.

2. Transporting dogs in vehicles is safer—and much less stressful for dog *and* human—when the dog rides well in a crate.

3. Crates are also useful if you—or your foster's new humans—stay in a hotel while travelling. Again, his own private bedroom goes with him, reducing his travel stress and ensuring you won't be paying for dog damage done to your hotel room.

4. Management—for you *and* for his new family—is much easier when he is happy about crating.

5. Housetraining also happens more easily when you can confine your foster to his crate in between bathroom trips.

Now we will turn to problem behaviors often associated with dogs who need foster care that require more significant training and management skills.

5

Problem Behavior: The Myth of Dominance and Alpha

This first behavior problem is actually one you won't encounter because it doesn't exist. However, for a number of reasons, you may think you are encountering it if your dog does not do what you want him to do. This is due to the fact that the dominance and alpha myth is everywhere. Google "alpha, dog" and you get more than 85 million hits. Really. While not all the sites are about dominating your dog, there are literally millions of resources out there—websites, books, blogs, television shows, veterinarians, trainers and behavior professionals—instructing you to use force and intimidation to overpower your dog into submission. They say that you, the human, must be the alpha. They're all wrong. Every single one of them.

Because the alpha/dominance dog myth is so pervasive in our culture today, a number of very normal "problem" behaviors that your foster dog may engage in at some point might lead you to think you have a dominant dog on your hand who needs to be put in his place. I have heard many common behaviors being attributed to a dog who is supposedly dominant including:

- Mounting other dogs
- Guarding his food bowl
- Housetraining accidents
- Refusing to bring a thrown ball back to his owner
- Growling when on leash when another dog gets close
- Putting his paw over your leg when you are sitting

In fact, "asserting dominance over" your foster dog every time you think he is misbehaving is one of the worst things you could possibly do. Trust me on this. His faith in humans may already be shaky because of his recent experiences. He needs you to repair his damaged trust, not hurt or frighten him further. Let's take a few pages here to explore why so many people buy into the dominance myth.

A history of dominance theory

The erroneous approach to canine social behavior known as **dominance theory** is based on a study of captive zoo wolves conducted in the 1930's and 1940's by Swiss animal behaviorist Rudolph Schenkel, in which the scientist concluded that wolves in a pack fight to gain dominance, and the winner is the alpha wolf. Schenkel's observations of captive wolf behavior were erroneously extrapolated to wild wolf behavior, and then to domestic dogs. It was postulated that wolves were in constant competition for higher rank in the hierarchy, and only the aggressive actions of the alpha male and female held the contenders in check. Other behaviorists, following Schenkel's lead, also studied captive wolves and confirmed his findings: groups of unrelated wolves brought together in artificial captive environments do, indeed, engage in often violent and bloody social struggles.

The problem is, that's not *normal* wolf behavior. As researcher David Mech stated in a 2002 article entitled "Alpha Status, Dominance and Division of Labor in Wolf Packs": "Attempting to apply information about the behavior of assemblages of unrelated captive wolves to the familial structure of natural packs has resulted in considerable confusion. Such an approach is analogous to trying to draw inferences about human family dynamics by studying humans in refugee camps. The concept of the alpha wolf as a 'top dog' ruling a group of similar-aged compatriots is particularly misleading."

*While captive wolves may engage in violent combat,
it is not normal wild wolf behavior.*

What we know now, thanks to Mech and others, is that in the wild a wolf pack is a family, consisting of a mated pair and their offspring of the past one to three years. Occasionally two or three families may group together. As the offspring mature, they disperse from the pack; the only long-term members of the group are the breeding pair. By contrast, in captivity unrelated wolves are forced to live together for many years, creating tension between mature adults that doesn't happen in a natural wild pack.

But that's all about wolves anyway, not dogs. How did it happen that dog owners and trainers started thinking all that information (and misinformation) about wolf behavior had anything to do with dogs and dog behavior? According to an article in the July 30, 2010 issue of *Time* magazine, somewhere along the line the logic went something like this: "Dogs are descended from wolves. Wolves live in hierarchical packs in which the aggressive alpha male rules over everyone else. Therefore, humans need to dominate their pet dogs to get them to behave."

Cesar Millan, the current darling of the dominance crowd, is only the latest in a long line of dominance-based trainers who advocate forceful techniques such as the alpha roll. Much of this style of training has roots dating back to the military of the early 1900s—which explains the emphasis on punishment. As far back as 1906, Colonel Konrad Most was using heavy-handed techniques to train dogs in the German army, then police and service dogs. He was joined by William Koehler after the end of World War II. Koehler also initially trained dogs for the military prior to his civilian dog training career, and his writings advocated techniques that including hanging and helicoptering a dog into submission (into unconsciousness, if necessary). To stop a dog from digging, he suggested filling the hole with water and submerging the dog's head in the water-filled hole until he was nearly drowned. If you try that today, I hope you get prosecuted for animal cruelty.

Fast-forward several years to 1978 and the emergence of the Monks of New Skete as the new model for dog training, asserting a philosophy that "understanding is the key to communication, compassion, and communion" with your dog (*How to be Your Dog's Best Friend*, Monks of New Skete; Little, Brown, 1978). The Monks were considered cutting edge at the time, and were in fact responsible for the widespread popularization of the "Alpha-Wolf Roll-Over" (now shortened to the alpha roll), in a complete and utter misinterpretation of the submissive roll-over that is *voluntarily offered* by the less assertive dog, not forcibly commanded by the stronger one. They also advocated the frequent use of other physical punishments such as the *scruff shake* (grab both sides of the dogs face and shake, lifting the dog off the ground) and *cuffing* under the dog's chin with an open hand several times, hard enough to cause the dog to yelp.

Even their most recent book, *Divine Canine; the Monks' Way to a Happy, Obedient Dog* (2007), while professing that "training dogs is about building a relationship that is based on respect and love and understanding," is still heavy on outdated, erroneous dominance theory. Immediately following their suggestion that "a kindly, gentle look tells the dog she is loved and accepted," they say, "But it is just as vital to communicate a stern reaction to bad behavior. A piercing, sustained stare into a dog's eyes tells her who's in charge; it establishes the proper hierarchy of dominance between person and pet." *(Author's note: It can also elicit a strong aggressive response if you choose the wrong dog as the subject for your piercing, sustained stare!)*

Enter the clicker

Just when it seemed that dog training had completely stagnated in turn-of-the-century military-style dominance-theory training, marine mammal trainer Karen Pryor wrote her seminal book, *Don't Shoot the Dog*. Published in 1985, this small, unassuming volume was intended as a self-help book for human behavior, the author never dreaming that her modest book, paired with a small plastic box that made a clicking sound, would launch a massive paradigm shift in the world of dog training and behavior. But it did.

Forward progress was slow until 1993, when veterinary behaviorist Dr. Ian Dunbar founded the Association of Pet Dog Trainers. Dunbar's vision of a forum for trainer education and networking has developed into an organization that now boasts nearly 6,000 members worldwide. While membership in the APDT is *not* restricted to positive reinforcement-based trainers, included in its guiding principles is this statement: "We promote the use of reward-based training methods, thereby minimizing the use of aversive techniques." The establishment of this forum facilitated the rapid spread of information in the dog training world, enhanced by the creation of an online discussion list where members could compare notes and offer support for a scientific and dog-friendly approach to training.

Enter the clicker, and the start of a massive paradigm shift in the dog training world.

Things were starting to look quite rosy for our dogs. The positive market mushroomed with books and videos from dozens of quality training and behavior professionals, including Jean Donaldson, Dr. Patricia McConnell, Dr. Karen Overall, Suzanne Hetts and others. With advances in positive training and an increasingly educated dog training profession embracing the science of behavior and learning and passing good information on to dog owners, at long last pain-causing, abusive methods such as the alpha roll, scruff shake, hanging, drowning and cuffing appeared to be headed the way of the passenger pigeon.

Then, in the fall of 2004, the National Geographic Channel launched its show, *The Dog Whisperer*, over the protests of several degreed behavior professionals to whom they had sent a review clip months earlier. Nat Geo was clearly informed in advance of that first airing that the star of the show, Cesar Millan, was using methods that were outdated, unscientific and potentially dangerous. "Don't do it," the experts warned. The show aired anyway. Dominance theory was back in vogue, with a vengeance. So despite the efforts of Donaldson, McConnell and the others noted above, today, everything from housetraining mistakes to jumping up to counter surfing to all forms of aggression is likely to be attributed to "dominance" by followers of the alpha-resurgence.

Why not alpha?

"But," some will argue, "look at all the dogs who have been successfully trained throughout the past century using the dominance model. Those trainers can't be *all* wrong."

In fact, harsh, force-based methods are a piece of operant conditioning (positive punishment), and as the decades have proven, those methods *can* work. They are especially good at *shutting down* behaviors—convincing a dog that it's not safe to do anything unless instructed to do something. And yes, that works with some dogs. With others, not so much.

My own personal, unscientific theory is that dog personalities lie on a continuum from very soft to very tough. Harsh, old-fashioned dominance-theory methods *can* effectively suppress behaviors without *obvious* fallout (although there is *always* behavioral fallout) with dogs nearest the center of the personality continuum—those who are resilient enough to withstand the punishment, but not so tough and assertive that they fight back. Under dominance theory, when a dog fights back, you must fight back harder until he submits, in order to assert yourself as the pack leader, or alpha. Problem is, sometimes they don't submit, and the level of violence escalates. Or they submit for the moment, but may erupt aggressively again the next time a human does something violent and inappropriate to them. Under dominance-theory training, those dogs are often deemed incorrigible, not suitable for the work they're being trained for nor safe as a family companion, and sentenced to death. Many of them, had they never been treated so inappropriately in the first place, might have been perfectly fine.

At the opposite end of the spectrum, a very soft dog can easily be psychologically damaged by one enthusiastic inappropriate assertion of rank by the heavy-handed dominance trainer. This dog quickly shuts down, fearful and mistrusting of the humans in his world who are unpredictably and unfairly violent.

Most crossover trainers (those who originally trained using old-fashioned methods and now are proud to promote positive reinforcement-based training) will tell you they successfully trained lots of dogs the old way. They loved their dogs and their dogs loved them. I'm a crossover trainer, and I know that's true. I also know that I would dearly love to be able to go back and redo all of that training, to be able to have an even better relationship with those dogs, to give them a less stressful life, and one filled with even more joy than the one we shared together. Where's a time machine when you need one?

Finally, the very presumption that our dogs would even consider us humans to be members of their canine pack is simply ludicrous. They know how impossibly inept we are, for the most part, at reading and understanding the subtleties of canine body language. We are equally inept, if not even more so, at trying to mimic those subtleties. Any attempts on our part to somehow insert ourselves into their social structure and communicate meaningfully with them in this manner are simply doomed to failure. It's about time we give up trying to be dogs in a dog pack and accept that we are humans co-existing with another species—and that we're most successful doing so when we co-exist peacefully.

Dogs know that humans are not dogs. They really do.

The fact is, successful social groups work because of voluntary deference, *not* because of aggressively enforced dominance. The whole *point* of social body language rituals is to *avoid* conflict and confrontation, not to *cause* it. Watch any group of dogs interacting. Time and time again you'll see dogs deferring to each other. It's not even always the same dog deferring:

Dog B: "Hey, I'd really like to go first."

Dog A: "By all means, be my guest."

Dog B passes down the narrow hallway.

Dog A: "I'd really like to have that bone."

Dog B: "Oh sure—I didn't feel like chewing right now anyway."

Dog A gets the bone.

What we know now is that yes, social hierarchies do exist in groups of domesticated dogs, and in many other species, including humans. We also know that a hierarchy can be fluid. As described above, one dog may be more assertive in one encounter, and more deferent in the next, depending on what's at stake, and how strongly each dog feels about the outcome. There are a myriad of subtleties about how those hierarchies work, and how the members of a social group communicate—in any species.

We also know that canine-human interactions are not driven by social rank, but rather by reinforcement. Behaviors that are reinforced repeat and strengthen. If your foster dog is doing an inappropriate behavior such as counter surfing or getting on the sofa, it's not because he's trying to take over the world—it's just because he's reinforced by finding food on the counter, or by being comfortable on the sofa. He's a scavenger and an opportunist, and the goods are there for the taking. Figure out how to prevent him from being reinforced for the behaviors you don't want, and reinforce him liberally for the ones you do. You'll be teaching him that he can trust the humans in his world and setting him up for success when he is finally placed in his lifelong loving home.

Social groups work best for all species courtesy of voluntary deference, not because of aggressively enforced dominance.

Quotes and resources on "alpha" dominance theory

There is a growing body of information available to anyone who wants to learn more about why dominance theory is so outdated and incorrect. Here are ten resources to get you started:

1. The American Veterinary Society of Animal Behavior Position Statement on Dominance (excerpt): "The AVSAB recommends that veterinarians not refer clients to trainers or behavior consultants who coach and advocate dominance hierarchy theory and the subsequent confrontational training that follows from it." (http://www.avsabonline.org/avsabonline/images/ stories/Position_Statements/dominance%20statement.pd)

2. The Association of Pet Dog Trainers Position Statement on Dominance (excerpt): "The APDT's position is that physical or psychological intimidation hinders effective training and damages the relationship between humans and dogs. Dogs thrive in an environment that provides them with clear structure and communication regarding appropriate behaviors, and one in which their need for mental and physical stimulation is addressed. The APDT advocates training dogs with an emphasis on rewarding desired behaviors and discouraging undesirable behaviors using clear and consistent instructions and avoiding psychological and physical intimidation. Techniques that create a confrontational relationship between dogs and humans are outdated." http://www.apdt. com/about/ps/dominance.aspx

3. Certified Applied Animal Behaviorist Kathy Sdao (article excerpt): "But even if dogs did form linear packs, there's no evidence to suggest that they perceive humans as part of their species-specific ranking. In general, humans lack the capability to even recognize, let alone replicate, the elegant subtleties of canine body language. So it's hard to imagine that

dogs could perceive us as pack members at all."
http://www.kathysdao.com/articles/Forget_About_
Being_Alpha_in_Your_Pack.html

4. Dr. Patricia McConnell, PhD—ethologist (article excerpt): "People who argue that ethology supports 'getting dominance over your dog' are not only focused on an issue more relevant 50 years ago than today, they are misrepresenting the findings of early researchers on social hierarchy. Social hierarchies are complicated things that allow animals to live together and resolve conflicts without having to use force every time a conflict comes up." http://www.4pawsu.com/pmdominance.htm

5. Dr. Meghan Herron, DVM (article excerpt): "Our study demonstrated that many confrontational training methods, whether staring down dogs, striking them, or intimidating them with physical manipulation such as alpha rolls [holding dogs on their back], do little to correct improper behavior and can elicit aggressive responses." http://www.sciencedirect.com/science/article/pii/S0168159108003717

6. Dr. Sophia Yin, DVM (blog excerpt): "Experts say dominance-based dog-training techniques made popular by TV can contribute to bites." http://drsophiayin.com/blog/entry/experts_say_dominance-based_dog_training_techniques_made_popular_by_televis

7. Study—University of Bristol (article excerpt): "Far from being helpful, the academics say, training approaches aimed at 'dominance reduction' vary from being worthless in treatment to being actually dangerous and likely to make behaviours worse." University of Bristol (2009, May 21). Using 'Dominance' To Explain Dog Behavior Is Old Hat. *ScienceDaily*. http://www.sciencedaily.com¬ /releases/2009/05/090521112711.htm

8. Jean Donaldson: *The Culture Clash* (James and Kenneth Publishing), 1996, 2005 "The dominance panacea is so out of proportion that entire schools of training are based on the premise that if you can just exert adequate dominance over the dog, everything else will fall into place. Not only does it mean that incredible amounts of abuse are going to be perpetrated against any given dog, probably exacerbating problems like unreliable recalls and biting, but the real issues, like well-executed conditioning and the provision of an adequate environment, are going to go unaddressed, resulting in a still-untrained dog, perpetuating the pointless dominance program."

9. Barry Eaton: *Dominance in Dogs; Fact or Fiction*, (softcover book) 2011, Dogwise Publishing: "...the alpha wolf is not the dictator of a pack, but a benevolent leader, and domestic dogs are not dictatorial and are unlikely to try to raise their status to rule over other dogs in a pack environment...I believe it's time to open our minds and consider the concept of pack rules as a thing of the past and recognize that dogs are not constantly trying to dominate their owners."

10. James O'Heare: *Dominance Theory and Dogs*, (softcover book) 2008, Dogwise Publishing: "...while the notion of social dominance holds potential for value in a social psychology and ethology context, it is an insidious idea with regards to explaining and changing behavior between companion dogs or dogs and people...it should be abandoned completely in that context in favor of a more efficient, effective and scientifically defensible behavioral approach."

From an APDT (Association of Pet Dog Trainers) interview with James O'Heare:

"The most significant problem with viewing dog-human relationships in the context of social dominance is that it implies and promotes an adversarial relationship between the two. It sets up a win-lose scenario, that actually ends up in a lose-lose scenario (as most win-lose scenarios do). It is incompatible with cooperation by its very nature, cooperation being something you need to promote an effective bond and training environment."

6

Common Problem Behaviors: Barking, Escaping, House Soiling

Some dogs who end up in shelters and rescues truly are perfectly well behaved. Many, through no fault of their own, come with one or more undesirable behaviors. Although "moving" is probably the most common reason offered for giving a dog up, the difference between "moving and taking my dogs with me, whatever it takes" and "moving and leaving the dog at a shelter" is often the unspoken "because he does thus-and-such and is a royal pain in the patootie." Sometimes the unwanted behaviors that cause a dog to lose his home are relatively simple, and might not even be a problem in a more tolerant, understanding or dog-savvy home. Sometimes they are significant and serious.

There are many good books on training and behavior modification, many of them quite detailed—and long. I'll touch on a few common problem behaviors associated with foster dogs here, address a couple of the more serious ones in the next few chapters, and encourage you to add some of the recommended books from the list at the end of this book to your library as resources to have on hand when a problem behavior presents itself in the form of your foster dog.

Barking

There's a lot of talk these days about the fact that dogs are primarily body language communicators. It's true, they are. But as anyone who's spent time with them knows, dogs also have a pretty well-developed ability to express themselves vocally. Dogs bark. Some bark more, some bark less, and a few don't bark at all, but most dogs bark at least some of the time. Most barking is actually normal behavior that's only considered inappropriate, or at least undesirable, by humans. Interestingly, while wild *puppies* may be quite vocal, wild *adult* dogs rarely bark, at least not to the degree our canine companions do. Genetics plays a large role here, of course. Over the millennia, we humans have been selectively breeding dogs and we've purposely selected for some breeds of dogs to be more vocal, others to be quiet.

At the "more" end of the continuum, the Scent Hounds are pro-grammed to give voice to announce the presence of their quarry. Thus Beagles, Coonhounds, Foxhounds and others in this group are quite vocal—although they do tend to bay rather than yap. Most of the Herding breeds are easily incited to bark. Skilled at telling a recalcitrant sheep or cow to back off, these Type-A workaholic dogs also delight in playing the role of noisy fun- police. Many of the Toy breeds also have a well-deserved reputation for barkiness, as do the Terriers. My three *very* barky dogs (out of five) are two Herding dogs and a Toy. Go figure.

In the "less barking" category, the Guarding breeds tend to reserve their formidable vocalizing for serious provocation. Sight Hounds also lean toward the quiet side, preferring to chase their quarry rather than bark at it. Then, of course, there's the Basenji—a somewhat primitive African breed of dog who doesn't bark—but he sure can scream!

Dogs bark for various reasons. If your foster dog arrives at your door with an unacceptable level of barking behavior, it's helpful to know why he's barking. Different barks have different tones; you can learn to recognize the differences in dog voices. Here are descriptions of the different kinds of barking, and what to do about them:

Alert/alarm barking. This is the dog who saves his family from a fire, tells us that Timmy's in the well, scares off the rapist, barks at the dogs on Animal Planet, and goes bonkers every time someone walks past on the sidewalk outside the picture window. Alarm barkers can save lives, but sometimes their judgment about what constitutes an alarm-appropriate situation can be a little faulty.

If your foster is alarm barking, reduce his exposure to the inciting stimuli. Perhaps you can baby gate him out of the front room, move the sofa away from the windows so he can't jump up and see out, or close the drapes. Outside, you might consider putting slats in the chain link fence to cut down on his visual access to the world surrounding his yard (better yet, install a privacy fence) or put up an interior fence to block access to the more stimulating parts of the yard. If major fence reconstruction isn't in your fostering budget, simply limit his access to the back yard to supervised visits, where you can interrupt his barking. Keep in mind, however, that your foster dog might be barking because something really *is* wrong. Take a moment to see what he is barking at. Perhaps your house really *is* on fire.

Demand barking. This behavior is more likely to annoy you than your neighbors, but it's annoying nonetheless, and prospective adopters may find it off-putting. A demand barker has learned that he can get what he wants, usually attention, or treats, by telling you. It often starts as a gentle, adorable little grumble, and can quickly turn into insistent, loud barks—your foster dog's way of saying, "I want it, NOW!"

Demand barking is easiest to extinguish early. The longer a dog successfully demands stuff, the more persistent he'll be if you try to ignore him. However, ignoring him is the best answer to this behavior. No treats, no attention—not even eye contact. The instant the demand behavior starts, utter a cheerful "Oops!" and turn your back on the dog. When he's quiet, say, "Quiet, yes!" and return your attention—and a treat—to him. Negative punishment—his barking makes a good thing, you, go away—works more quickly when followed by positive reinforcement for a desirable behavior: quiet brings you—and your attention—back.

Watch out for extinction bursts and behavior chains. When you're trying to make a behavior go away by ignoring it, the dog may increase the intensity of his behavior—"I WANT IT NOW!" This is an **extinction burst.** If you succumb, thinking it's not working, you reinforce the more intense behavior, and your foster is likely to get more intense, sooner, the next time. If you stick it out and wait for the barking to stop, you're well on your way to making it go away. You have to be more persistent—and consistent—than the dog. A **behavior chain** is a series of behaviors linked together. In order to avoid the behavior chain of bark-quiet, bark-quiet, bark-quiet, vary the amount of time you have him be quiet before returning your attention to him. Short periods of time at first, then gradually—and randomly—longer periods interspersed with shorter ones.

Frustration/arousal barking. Often confused with anxiety barkers, dogs who have a low tolerance for frustration will bark hysterically when they can't get what they want. Unlike a separation anxiety panic attack, this is simply an "I WANT IT!" style temper tantrum similar to demand barking, but with more emotion, and directed at the thing she wants, such as a cat strolling by, rather than at you (although it may be you she wants, as well). If you consistently offer high-value treats in the presence of frustration-causing stimuli (preferably *before* the barking starts), you can counter condition your dog to look to you for treats when the cat strolls by (cat = yummy treats) rather than erupt into a barking fit.

Boredom barking. This is the dog who's left out in the back yard all day, perhaps on a chain, or maybe all night. Dogs are social creatures, and the back yard dog is lonely and bored. Boredom barking is often continuous, with a monotonous quality: "Ho hum, nothing else to do, I may as well just bark." This is the kind of barking that's most annoying to neighbors, and most likely to elicit a knock on your door from a friendly animal services officer.

The answer here is obvious, and relatively easy. Bring the dog inside. In fact, I would fervently hope you aren't even planning to leave your foster dog outside. Many outdoor barkers are perfectly content to lie quietly around the house, indoors, waiting for you to come home, and sleep peacefully beside your bed at night. If your foster dog isn't house-safe, use crates, exercise pens, dog walkers, lots of exercise, even doggie daycare to keep him out of trouble, until he earns house privileges. You can also enrich his environment by giving him interactive toys such as food-stuffed Kongs that keep his brain engaged and his mouth busy.

Stress barking. Stress barkers are fearful, anxious or even panicked about something real or anticipated in the environment, such as the actual or possible approach of a threat, or isolation distress/separation anxiety. Separation anxiety is manifested in a number of behaviors, including non-stop hysterical barking and sometimes howling. If your foster is barking due to stress, fear or anxiety, consult with a qualified professional behavior counselor who uses positive reinforcement-based methods, and try to manage his environment to minimize his exposure to stressors while you work on a behavior modification program. It will be important to make good progress with modifying this behavior prior to placement, or he is likely to be returned by his adopters.

Play barking. This is a common behavior for herding dogs—the cheerleaders and fun- police of the canine world. As other dogs—or humans—romp and play, the play-barker runs around the edges, barking, sometimes nipping heels.

If you're in a location where neighbors won't complain and the other dogs tolerate the behavior, you might just leave this one alone. With children, however, the behavior's not appropriate, and the dog should be managed by removing him from the play area, rather than risk injury to a child.

If you do want to modify play-barking behavior, use negative punishment—where the dog's behavior makes the good stuff go away. When the barking starts, use a loss-of-opportunity marker such as "Oops! Too bad!" and gently remove your foster from the playground for one to three minutes. A tab—an old leash cut down to six to twelve inches left attached to his collar—makes this maneuver easier. Then release him to play again. Over time, as he realizes that barking ends the fun, he may start to get the idea. Or he may not—this is a pretty strong genetic behavior, especially with the herding breeds. You may just resort to finding appropriate times when you allow play-barking to happen.

You need to figure out why your foster dog is barking in order to effectively manage and modify the behavior.

Uncontrolled barking can be frustrating to humans. I know this all too well, with several vocal dogs in our own personal pack. However, our dogs sometimes have important and interesting things to say. Don't lose sight of the value of their vocal communications—they may be trying to tell you something important. Learn the tones—they are talking to you! If you ignore them the house might burn to the ground, or Timmy might really drown in the well.

Escaping

If your foster dog was found as a stray, he may have been dumped by his human, but there's at least an equal chance he's just good at escaping. Dogs escape because they can. It might be something as simple as a past owner who opened the door to an unfenced yard to allow the dog to relieve himself (which I know *you* won't do) or as challenging as the escape artist who's dedicated to darting out doors and jumping over, digging under or chewing through fences. Boredom, arousal and fear are common motives for dogs to escape, as well as the reinforcement they get when they discover how much fun it is to roam the neighborhood, chase bunnies and squirrels, get in garbage, play with other dogs or the kids next door, pursue females in season, and get handouts from dog-loving humans.

Prevention and management are critically important to keeping your foster safe in his temporary new home. The last thing you want to do is leave him in the back yard for even a brief time, and return to find him *gone*.

*This fence may be adequate to contain your "home dog,"
but is far too flimsy to trust to your fosters.*

You have a "fresh-start advantage" with your foster dog. Even though he may have been quite the accomplished escaper in his old life, if you're just bringing him home he has *never* escaped from *yours*. Don't let him learn he can escape from his new environment—stop any embryonic escape attempts in their tracks by taking the following prophylactic measures:

1. **Provide a safe, secure enclosure.** Before your foster comes home, make sure your fence is flush to the ground, or even buried a few inches. Check for rotten spots, and crawl behind shrubs and brush to look for holes or loose boards. You *know* the dog will find them if you don't! If there are vulnerable spots that you cannot bury, reinforce them with heavy cement blocks.

2. **Go overboard on fence height.** Raise the fence to at least five feet for a small dog (perhaps higher for very athletic small dogs like Jack Russell Terriers) and six feet for medium to large dogs. Make sure there are no woodpiles, doghouses, deck railings or other objects close enough to the fence to provide a convenient launch pad. If you can't afford to raise and reinforce the fence, don't ever leave your foster alone in the yard.

3. **Teach your foster to wait at doors until invited through.** (This lesson will serve him well in his new home, as well.) Use "Wait" at every door to the outside world, every time you open it, whether the dog is going to go through or not.

4. **Install dog-proof latches on gates and springs to ensure they always close.** There's no point in waiting until after your foster is hit by a car, or vanishes into the unknown, to discover that he can work the latch. In fact, a padlock will prevent accidental release from the outside by a visitor or intruder at the same time it keeps your foster from practicing his latch-opening skills.

5. **Keep your foster dog indoors or in a very secure kennel when you are not home.** Boredom and loneliness are strong escape-motivators, and he has plenty of time to plan and execute the great escape when you are not there to interrupt unwanted behaviors such as digging under and chewing through fences.

6. **Remember those ID tags!** Just in case he manages to escape despite all your precautions...

Teaching "Wait"

The goal with Wait exercises is to help your foster dog learn to remain still before you let him know it is okay to move again. Your dog will succeed if you shape the behavior in small steps. Any time the dog stops succeeding, you've made the steps too big, or tried to take too many steps too quickly. Always seek to find the place where the dog wins, and move forward from that place in tiny steps.

Wait at the door. With your foster dog sitting at your side at a door that opens *away* from you, tell him to "Wait." Reach your hand a few inches toward the doorknob. If the dog doesn't move, click your clicker or use your verbal marker, and feed him a tasty treat. Repeat this step several times, moving your hand closer toward the doorknob in small increments, clicking and treating each time he stays sitting.

Remember that you're shaping the behavior in tiny steps. If he gets up, say "Oops!" and have him sit, then try again. If he gets up two or more times in a row, you're advancing too quickly; go back to moving your hand only a few inches toward the knob, and make your increments even smaller.

When he'll stay sitting as you move your hand toward the door, try actually touching the knob. Click and treat. Then jiggle the door knob. Click and reward for not moving. Repeat several times, clicking and treating each time, then slowly open the door a crack. If the dog doesn't move, click and treat. If he does get up, say "Oops!" and close the door. You're teaching him that getting up makes the door close—if he wants the possible opportunity to go out, he needs to keep sitting.

Gradually open the door a bit more, an inch or two at a time. Any time he gets up, "Oops!" and close the door, and try again. If you get two or three "Oopses" in a row you're doing too much—back up a few steps and progress more slowly. Click and reward for not moving, several times at each step. When you can open the door all the way, take one step through it, stop, turn around and face the dog. Wait a few seconds, click, then return to the dog and give a food reward.

When he's really solid with you walking out the door, you can sometimes invite him to go out the door ahead of you, with you or after you, and sometimes walk through and close the door, leaving him inside as you would if you were leaving for work. Once the door has closed, he's free to get up and move around.

One of the wonderful things about the "Wait" cue is that dogs do seem to generalize it pretty easily. If you teach it at a door in your home, they'll understand pretty quickly when you ask them to "Wait!" when you open the car door—a great safety behavior so your foster dog doesn't jump out on the highway if you have to get out of the car on the side of the road to change a flat tire. Once you've taught Wait with the food bowl and door, try it on a leash walk. If the dog starts to move too far out in front of you, say "Wait!" If he doesn't pause of his own accord, stop moving and the leash will stop him (*don't* jerk him to a stop!). A few repetitions of this and he'll figure it out in no time. Then, when he gets adopted, remember to show his new humans this impressive and useful behavior.

Wait for food. *Note: If your foster dog guards his food bowl aggressively, don't teach this exercise until you have successfully modified the resource guarding behavior.* "Wait" can also be used to avoid mad scrambles at feeding time. With your foster dog sitting at your side, hold the bowl at chest level and tell him to "Wait." Move the food bowl (with food in it, topped with tasty treats) toward the floor four to six inches. If the dog stays sitting, click your clicker (or use a verbal marker such as "Yes," or a tongue click) and feed him a treat from the bowl. If the dog gets up say "Oops," raise the bowl back up and ask him to sit again. If he remains sitting, lower the bowl four to six inches again, Click and treat if he's still sitting.

If he gets up a second time, say "Oops," raise the bowl and have him sit. On your next try, only lower the bowl an inch or two. Click and treat. Repeat this step several times until he consistently remains sitting as you lower the bowl. Gradually move the bowl closer to the floor with successive repetitions until you can place it on the floor without the dog trying to get up or eat it. After each repetition, stand up straight and raise the bowl all the way back up.

Finally, place the bowl on the floor and tell your foster dog to eat. After he has had a few bites, lift the bowl up and try again. Repeat these steps until you can place the bowl on the floor and he doesn't move until you tell him to. One of the great things about the food bowl Wait is that if you feed your foster dog twice a day, you already have two natural training sessions built into your schedule!

Housetraining

Poop and pee outside, not inside. It seems like such a simple concept, and indeed, many dogs are easily housetrained. But not all. We adopted our Scorgidooodle (Scottie/Corgi/Poodle), Bonnie, when she was about six months old. She was surrendered to the shelter because her owners couldn't housetrain her. By age four she was *reasonably* reliable with her housetraining, but she'll never be one of those dogs you'll find waiting at the door with legs crossed if an emergency keeps you away from home for ten hours.

Housetraining is a critically important skill to teach your foster, if it's one he doesn't come with. Housesoiling is a common reason for adoption returns. You want your foster to *stay* in his next-and-forever home.

The basic premise of successful housetraining is: Take the dog out more often than he has to go, and supervise and manage to prevent accidents. For a young puppy, that means outside every-hour-on-the-hour during the day, and crated or otherwise confined at night, where you can hear him if he wakes up and says he has to go out. As the pup matures, a general rule of thumb is that he can "hold it" for one hour more than his age in months. That is, a three-month-old pup might be able to go for four hours—up to a maximum of about eight hours on a regular basis for an adult dog. Remember, though, that it's a general rule—there are certainly normal three-month-old puppies who can only hold it for two to three hours and adult dogs who can't go longer than three or four. There are also canine saints who will maintain their housetraining when left alone for ten or even twelve hours—but even with those I don't recommend expecting them to do so routinely.

The basis for successful housetraining: Take your foster dog out to relieve himself more frequently than he has to go.

Alternatively, if you are fostering a litter of young pups, you can house them in an exercise pen set on top of a waterproof tarp, covered with a thick pad of newspapers. They will learn to eliminate on a newspaper substrate, which will suffice until they are old enough and in an environment where housetraining can happen. If you foster them after they reach the age of eight weeks, begin their housetraining by confining them to a crate for part of the time, and taking them from the crate immediately outdoors, so they start learning to eliminate on a grass or gravel substrate.

We rely on a dog's instinct to keep the den clean as the basis for housetraining. Your "den" (your home) is huge compared to a dog's natural den, so your foster, until he learns otherwise, sees nothing wrong with using part of your vast empire as his bathroom. A crate appropriate to your foster's size—big enough to stand up, turn around and lie down in, is a perfect den. He won't want to soil it, he will let you know when he has to go out, and he'll learn how to "hold it." Of course, you still have to house train—learning to "hold it" in the crate isn't the same as learning to "hold it" in your whole house.

Crates can be wonderful tools. Properly used, your foster learns to love the crate; it's his own comfortable, safe space, and it can go with him to his new home, which can help ease the stress of adoption. Crates can also cause problems. "Over-crating" is the abusive use of a crate—keeping a dog in his portable den for longer than is physically and/or mentally healthy for him. Some people do it out of ignorance, not realizing that it's not okay to crate a dog ten hours a day, five days a week, while they're at work. If I had to be gone that long every day I'd arrange for a pet sitter to come in and let my dogs out for a noontime potty break—crates or no crates! Ten hours is just too long for your foster to be expected to consistently wait to relieve himself. Of course there are dogs who can do it, and every once in a while when something prevents you from getting home in time it may be no big deal, but it's not a good regular practice.

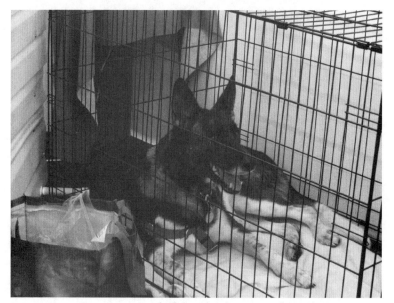

Crates are wonderful tools when properly used.
Improperly used, they are torture devices.

Some people deliberately overuse crates. These are the worst abusers. Puppy mills, hoarders, misguided rescuers and foster homes (not the good ones!), and uncaring or uneducated owners may routinely crate dogs anywhere from ten to as much as 24 hours a day. This is abuse, pure and simple. These dogs aren't getting the exercise and social interaction they need to be physically and mentally healthy, which contributes significantly to behavior problems if and when the dogs are fortunate enough to be rescued and adopted. In addition, their natural instincts to keep their dens clean are totally destroyed. These can be the most challenging dogs to housetrain, since they will willingly urinate and defecate in their crates. It's all they know.

I recommend the **umbilical approach** to housetraining puppies and adult dogs. This means that your foster dog is always either in a crate or pen, on a leash held in your hand or clipped to your belt (or restrained nearby on a tether, or in the same room with the door closed), under the direct supervision of an adult or responsible teen, or outdoors, until he can be trusted with house freedom.

Establish a daytime routine—go out *with* your foster every one to two hours. Don't just send him out to "do his business" on his own. You won't know if he did anything or not, and you won't be able to reward him for doing the right thing. Go with him. When he urinates or defecates tell him "Yes!" and feed him a treat. Be sure to wait until he's just about empty—if you interrupt his flow you may think he's empty when he's not. Then play with him for a few minutes before bringing him indoors, as an additional reward for eliminating. *If you bring him back in immediately after he eliminates, he may learn to "hold it" as long as possible when you take him out, in order to prolong the outdoor excursion.* If he doesn't go, bring him back in, put him in his crate, and try again in a half hour or so. When you know he's empty you can give him some relative-but-still-supervised freedom for twenty minutes to a half hour. Then he goes back under wraps.

If he makes a mistake indoors, do *not* punish him after the fact. It's your mistake, not his. He won't even know what he is being punished for! Quietly clean it up, using an enzyme-based cleaner designed for clean up of pet waste, and vow not to give him so much freedom. If you must spank someone with a rolled up newspaper, hit yourself in the head three times while repeating, "I will watch the dog more closely; I will watch the dog more closely; I will watch the dog more closely."

If you catch him in the act, calmly interrupt with a cheerful, "Oops! Outside!" and take him outside. Again, do *not* punish him. If you do, you'll only teach him that it isn't safe to toilet in front of you; he'll learn to run to the back bedroom to do it.

If your dog is quicker than you, or has housesoiling challenges that include marking and excitement or submissive urination, a belly band (for males) or panties (for females) are useful management tools. You still need to make the effort to housetrain, but at least your carpets will be safe!

At night your foster dog should be confined, if possible crated in or near your bedroom. If he wakes up in the middle of the night and cries, he probably has to go out. You must wake up and take him out, click and reward when he eliminates, then bring him back and immediately return him to his crate. You don't want to teach him that crying at night earns a play session, or a snuggle in bed with you. (This is an exception to the above-stated "let him run around and play a bit after he eliminates" rule.)

Of course there are other possible inappropriate behaviors your foster dog may offer—far too many to include in this small book. Topping the list of more difficult behaviors you may encounter are fear and aggression. In fact, the most common presentation of aggression behavior professionals see is, in fact, fear-related aggression, not "dominance aggression" as some would have you believe. Both fear and aggression are behaviors not to be taken on by the faint of heart, or the novice dog person, even the very well-meaning novice dog person. We will address these in the chapters that follow, as well as a third behavior not uncommon to shelter and rescue dogs: separation/isolation distress. If you believe you are seeing any of these with your foster, *immediately* notify your foster organization and seek *qualified* professional help.

7

Problem Behavior: Helping Your Fearful Foster Dog Find Courage

President Franklin D. Roosevelt, speaking about the Great Depression, said, "We have nothing to fear but fear itself." If only it were that simple when dealing with dog behavior instead of failing economies. Fear-related behaviors can be debilitating to the inappropriately fearful dog, and many foster dogs exhibit these behaviors. They are heartbreaking, frustrating, even sometimes dangerous for the human trying to deal with her dog's strong emotional responses, and for the dog who may injure himself or others in his desperate efforts to escape or protect himself from the fear-causing stimulus. With the increase in dogs from hoarder and abuse cases being rescued, a great number of poorly socialized dogs with fear-related behaviors are ending up in shelters, rescue groups, and eventually foster homes. While some of these dogs rehabilitate with surprising ease, others are devastatingly difficult. I recommend you make a well-considered conscious decision about whether you want your foster experience to include working with fearful dogs.

Three fear-related behaviors

The complex of fear-related behaviors includes fears, anxieties and phobias. While they are closely related emotional responses, they differ significantly in several ways, including the presence or absence of a physical trigger, the intensity of the dog's response, and the ease with which the emotional response and related behaviors can be modified. In general, these three conditions can be among the most difficult behavioral problems to treat.

There is a strong genetic component to fear-related behaviors. Whereas once we tended to place a lot of the blame on owners for their perceived role in creating fearful dogs, today we recognize that a genetic *propensity* toward fearfulness is a significant factor in the actual manifestation of fear-related behaviors. While environment—especially lack of socialization—can play a critically important role in bringing these behaviors to fruition, genes explain why two dogs with similar upbringing and socialization can react so differently in the presence of a potentially fear-causing stimulus, and why even a well-socialized dog can suddenly develop phobic behaviors.

Fear is defined as a feeling of apprehension associated with the *presence or proximity* of an object, individual or social situation. It's a valuable, adaptive emotion, necessary for survival and appropriate in many situations. It's good to be afraid of grizzly bears, tornados and semi-trucks skidding out of control on icy highways. Your dog is wise to fear the flashing heels of a galloping horse, strong waves crashing on an ocean beach, the spinning wheels of a passing car. People and animals who feel no fear are destined to live short lives. As we saw in Chapter 3, dogs who were not properly socialized and exposed to a variety of stimuli during the critical period of the first 16 weeks of life are more likely to develop fear-based problems.

Of course, overly fearful dogs may lead short lives as well. Fear-related aggression is a significant risk to a dog's long and happy life. A fearful dog's first choice is usually to escape, but he may bite defensively if cornered or trapped, and dogs who bite are often euthanized. In addition, a constant emotional state of fear makes for a poor quality of life for a dog, and for humans who are stressed by their fearful dog's behavior.

Debates about anthropomorphism aside, most biologists agree that human and nonhuman mammals experience fear similarly. Recall one of your own heart-stopping, adrenaline-pumping life experiences. Perhaps you were approached by a menacing stranger in an alley on a dark night, threatened by a large predator on a camping trip, cornered by an angry bull in a pasture, or just missed rear-ending a car in front of you when a moment of inattention caused you to miss the warning flash of brake lights. Remember how helpless, vulnerable and terrified you felt? You can empathize with your dog when you see him trembling in the presence of a stimulus that elicits a similar response in his canine brain and body.

This shelter dog is fearful of the cats on the other side of the glass window.

Anxiety is the distress or uneasiness of mind caused by apprehensive *anticipation* of *future* danger or misfortune, real or imagined. Anxious dogs appear tense, braced for a threat they can't adequately predict, sometimes one that doesn't actually even exist. Anxiety can be a chronic condition, one that significantly impairs a dog's (and owner's) quality of life, and one that can be more challenging to modify than the fear of a real and present danger. Separation distress is perhaps the most widely discussed anxiety-related behavior in dogs, but owner absence is not the only cause for canine apprehension. Many dogs are anxious on car rides—anticipating, perhaps, a visit to the vet's office, or some other "bad" place. A dog who has been attacked by a loose dog while walking on leash may become anxious about going for walks, constantly stressed, scanning the neighborhood for another potential attacker.

Again, human anxieties are similar to canine. If you've been mugged in a dark alley, you are likely to experience some degree of stress anytime you find yourself walking down an alley in the dark. Some people experience extreme anxiety over taking exams, even when their past successes show that they pass tests with flying colors. Barbra Streisand, successful singer that she is, suffers from extreme performance anxiety, still becoming physically ill every time she's about to walk on stage. The danger or misfortune may be imagined, but the anxiety is very real.

Phobias are persistent, extreme, inappropriate fear or anxiety responses, far out of proportion to the level or nature of threat presented. They are stubbornly resistant to modification through habituation or desensitization (repeated low-level exposure to the stimulus that causes the extreme response). While inappropriate in degree, a phobic response is not totally irrational—it is usually directed toward something that *could* be harmful. Common human phobias are related to snakes, spiders, high places, flying—all things that have *potential* to be life threatening. In reality, the majority of snakes and spiders are relatively harmless, it's rare for humans to accidentally nosedive off a skyscraper, and only a tiny percentage of airplanes ever crash. Common canine phobias include extreme reactions to thunderstorms and other sounds, fear of humans, and **neophobia**—inappropriate and strong fear response to novel stimuli (anything new and different).

The face of fear

Most people, even non-dog owners, can identify a dog who is in abject fear: trembling, drooling, crouched low, tail tucked, pupils dilated, perhaps even losing control of bladder and bowels. It's much easier to miss the more subtle early warning signs indicating the initial onset of fear. Yet, as with all undesirable behaviors, fear is easier to deal with sooner, rather than later, so there is real value in being able to determine when your foster dog is slightly fearful and take prompt steps to alleviate the fear—by removing the fear-causing stimulus and/or implementing a program of counter conditioning and desensitization.

Ignorance of subtle fear signals is one of the primary reasons a purportedly "child-friendly" dog mauls the unsuspecting toddler. Because the dog never exhibited overt aggression to the child—growling, lunging, snapping—the owner assumed the dog was kid-friendly. Instead, the dog may have always felt threatened by the presence of children, with their high-pitched voices, sudden movements and sometimes inappropriate behaviors toward the dog. Misinterpretation of fear signals can also be the cause of inappropriate human behavior—punishing the dog or forcing the dog to confront the threat—both of which can worsen a dog's fears and trigger an aggressive response. To avoid exacerbating your foster dog's fear, or perhaps turning a fear into a phobia, watch for the following early warning signs and be prepared to protect your foster from the perceived threat:

- Making an effort to leave

- Hiding behind you

- Averting eyes

- Panting—increase in respiratory rate

- Sweaty paws—leaving footprints on cement

- Reluctance to take/eat treats

- Ducking the head

- Licking, yawning, blinking

Getting brave

Whether you're working with fears, anxieties or phobias, one effective solution to an inappropriate emotional response is counter conditioning and desensitization (CC&D) to *change* your dog's emotional response to the stimulus or situation. In *The Cautious Canine*, author and behaviorist Dr. Patricia McConnell calls counter conditioning a "universally effective treatment for fear-based behavior problems." Think of it as training your dog's emotions rather than training his actions. Behavior change will follow emotional change.

Counter conditioning a new emotional response

Counter conditioning involves changing your dog's association with a scary stimulus from negative to positive. The easiest way to give most dogs a positive association is with very high-value, really yummy treats. I like to use chicken—canned, baked or boiled—since most dogs love chicken and it's a low fat, low calorie food. Perhaps your dog is afraid of your vacuum cleaner. Here's how the CC&D process works:

1. **Determine the distance** at which your dog can look at the non-running, stationary vacuum cleaner, and be alert and a little wary but not extremely fearful. This is called the *threshold distance*.

2. **With you holding your dog on leash**, have a helper present the non-running vacuum at threshold distance X. The instant your dog sees the vacuum, start feeding bits of chicken, non-stop.

3. **After several seconds**, have the helper remove the vacuum, and stop feeding chicken.

4. **Keep repeating steps one through three** until the presentation of the vacuum at that distance consistently causes your dog to look at you with a happy smile and a "Yay! Where's my chicken?" expression. This is a *conditioned emotional response* (CER)—your dog's association with a non-running vacuum at threshold distance X is now positive instead of negative.

5. **Increase the intensity of the stimulus**. You can do this by decreasing distance to X minus Y, by increasing movement of the vacuum at distance X, or by turning the vacuum on. I'd suggest decreasing distance first in small increments by moving the dog closer to the location where the vacuum will appear, achieving your CER at each new distance, until your dog is happy to be right next to the non-running, non-moving vacuum, perhaps even sniffing or targeting to it.

6. **Then return to distance X and add movement** of your non-running vacuum, gradually decreasing distance and attaining CERs along the way, until your dog is delighted to have the non-running, moving vacuum in close proximity.

7. **Now, back to distance X, with no movement,** have your helper turn the vacuum on briefly, feed treats the instant it's on, then turn it off and stop the treats.

8. **Repeat until you have the CER, then gradually increase the length of time** you leave the vacuum running, until he's happy to have it on continuously.

9. **Begin decreasing distance in small increments,** moving the dog closer to the vacuum, obtaining your CER consistently at each new distance.

10. **When your dog is happy to have the running, stationary vacuum close to him,** you're ready for the final phase. Return to distance X and obtain your CER there, with a running, moving vacuum. Then gradually decrease distance until your dog is happy to be in the presence of your running, moving vacuum cleaner. He now thinks the vacuum is a *very good* thing, as a reliable predictor of very yummy treats.

The example in the above box concerns a fairly simple fear behavior. The more complex the stimulus, the greater the number of fear-causing stimuli; the more intense the response, the more challenging the behavior is to modify. Anxieties and phobias generally require a greater commitment to a longer term and more in-depth modification program, and often require the intervention of a good, positive behavior professional.

Get creative as you search for ways to help your fearful foster. Television programs offer ideal opportunities for counter conditioning and desensitization to various auditory stimuli, as do recordings of thunderstorms and applause, where the intensity of stimulus (volume) can be controlled. Real thunderstorms are another story, however. They are almost always super-threshold—occurring at an intensity that triggers a strong emotional response, trembling and shutting down to a degree where the dog can no longer accept high-value treats.

Consider other measures such as melatonin (check with your veterinarian), a snug T-shirt (the economy version of an Anxiety Wrap™: http://www.anxietywrap.com or Thundershirt www.thundershirt.com), which operates on the concept of "swaddling" as a comforting device, a Comfort Zone/Adaptil ™ plug-in DAP diffuser, and the use of anti-anxiety drugs obtained through consultation with a behavior-knowledgeable veterinarian. (Note: Although many vets prescribe it, Ace Promazine is usually the totally *wrong* drug for treating storm-related anxieties. If you are considering medication for any behavioral problem, it is critically important to work with a *behavior knowledgeable* veterinarian.) A Storm Defender™ cape (http://www.stormdefender.com) may be more effective than a snug T-shirt, and a Calming Cap™ (www.thundershirt.com) can help to reduce the intensity of the stimulus of lightning flashes.

Behavior modification drugs may also be indicated, but must be used in conjunction with a solid behavior modification program. Drugs alone won't fix your foster's fear-related behaviors. If you think drugs might be warranted, talk with your partner organization first, and then insist that your—or the group's—veterinarian do a phone consult with a veterinary behaviorist (VB). Most VBs will do phone consults with other veterinarians at no charge.

Courage-enhancing activities

In addition to early socialization for puppies, and counter conditioning for existing fears in puppies and adult dogs, there are a number of things you can do to help your foster dog, regardless of his age, to get brave. Some of these things, such as basic good manners training, help him understand his world and make communication flow more smoothly between dog and human which helps reduce stress at both ends of the leash. Others are games and activities you can do to help build his confidence. The more of these you can do with your timid or fearful foster, the better his chances are for a long and happy life. Be sure to let his new humans know which ones are his favorites!

Basic good manners training. You don't have to do a lot of fancy stuff to help your foster dog become more confident in his world. Simply teaching him basic good manners—to respond appropriately to your cues—will make his environment more predictable, because he knows what humans are asking of him, and he knows the consequences of his behavior. Of course it goes without saying that you will use positive reinforcement-based training with him so the consequences are happy ones. My book, *The Power of Positive Dog Training*, can help you do just that. Nothing can destroy a timid dog's confidence faster than the application of verbal or physical punishment that serves to convince him he's right to think the world is a scary and unpredictable place.

Combine your foster's positive reinforcement good manners training with structure in his routine and stability in his life and you will have taken a large step toward increasing his confidence. But of course, you want to do more to help him get brave. Happily, you can do that simply by doing fun stuff with him. Read on…

Basic good manners training is one of the best things you can do for your foster dog, to give him a leg-up in his new life. Be sure to use positive reinforcement-based methods!

Targeting. Targeting means teaching your dog to touch a designated body part to a designated target. That description doesn't do it justice…targeting is tons of fun! Many dogs *love* targeting, partly because it's easy to do, and partly because it pays off well—"push the button (the target spot), get a treat."

Since dogs naturally explore the world with their noses and paws, nose and foot targeting are the two easiest. Nose targeting draws your dog's eye contact and attention from a worrisome stimulus to a pleasant one, so that's the one I find most useful for timid dogs, although foot targeting can work too.

It's an embarrassingly simple behavior to teach. Hold out your hand in front of your foster, nose level or below. When he sniffs or licks it (because he's curious!), click your clicker and feed him a treat (or use a verbal marker—a mouth click, or a word—and treat). Remove your hand, then offer it again. Each time he sniffs, click and treat. If he stops sniffing (boring—I've already sniffed that!) rub a little tasty treat smell on it, and try again.

You're looking for that wonderful "light bulb" moment when he realizes he can make you click by bumping his nose into your hand. His "touch" behavior becomes deliberate, rather than incidental to the smells on your hand. When you see him *deliberately* bumping his nose into your hand, add the "Touch!" cue as you offer your hand to him. Encourage him with praise, and high-value treats. Make it a game, so he thinks it's the most fun in the world. You want to see his eyes light up when you say "Touch," and you want him to "bonk" his nose into your hand, hard! Start offering your hand in different places so he has to move to touch it, climb on something to touch it, jump up to touch it.

Now try playing the Touch game when your foster boy's a little bit nervous about something. Scary man with a beard and sunglasses passing by on the sidewalk? Hold out your hand and say "Touch!" Your dog takes his eyes—and his brain—away from the scary thing and happily bonks his nose into your hand. Click and treat. He can't be afraid of the man and happy about touching your hand at the same time. He also can't look at your target hand and stare at the scary man at the same time.

Ask him to "Touch" several more times, until the man has passed, and then continue on your walk. If you do this *every* time he sees a scary man, he'll decide that men with beards and sunglasses are good because they make the Touch game happen! By changing your dog's behavior—having him do something he loves rather than act fearful—you can manage a scary encounter, and eventually change his emotional response to and association with something previously scary to him. He's getting braver!

Find It. Like targeting, Find It is a behavior many dogs learn to love, and another game you can play to change behavior in the presence of a fear-causing stimulus, eventually changing your foster's emotional response. This is also another ridiculously easy and delightful game that *any* dog can play.

Start with the dog in front of you and handful of tasty treats behind your back. Say "Find It!" in a cheerful tone of voice and toss one treat a few feet to your left. When your foster gets to the treat, click just before he eats it. When he comes back to you say "Find It!" again and toss a second treat a few feet to your right. Click—and he eats the treat. Do this back and forth, until he is easily moving from one Find It treat to the other. Then toss them farther each time until your foster dog happily runs back and forth. Fun game—and exercise, to boot!

Now if a scary skateboarder appears while you're walking your foster dog around the block on his leash, play Find It, keeping the tossed treats close to you. Your dog will take his eyes off the scary thing and switch into happy-treat mode. You've changed his emotions by changing his behavior.

Targeting and Find It can also work to walk your timid foster past a scary, stationary object, like a manhole cover, or a noisy air conditioning unit. Touch-and-treat as you walk past, or toss Find It treats on the ground ahead of you and slightly away from the scary thing, to keep him moving happily forward.

Emergency Escape. Your Emergency Escape game gives you a "run away" strategy when you know an approaching stimulus will be too much for your worried foster. However, because you've taught it to him as a fun game, he's not running away in panic—he's just playing one of his favorite Get Brave games that just happens to move him farther away from the scary thing.

Teach this game to your foster dog in a safe, comfortable environment when he's not being fearful of something. As you are walking with him on leash, say your "Run Away!" cue, then turn around and run fast, encouraging him to romp after you for a squeaky toy, a ball, a handful of high-value treats at the end of the run, or a rousing game of tug—whatever your foster loves. The key to success with this exercise is convincing your foster dog that the "Run Away" cue is the predictor of wonderful fun and games. Again—you're teaching him a new, fun behavior—Run Away!—that you can use to change his emotional response in a scary moment. He's getting braver.

Teach your fearful foster dog an Emergency Escape as a fun game, and then use it to get him out of trouble when a fear-causing stimulus is too close.

Play. In fact, you can use any behavior your foster dog already loves—a trick, a toy, a game—anything that lights up his face—to convince him that good things happen in the presence of something scary. If he loves to roll over, ask him to do that. If he delights in snagging tossed treats out of the air, do that. High Five? Crawl? Spin and Twirl? Do those.

Invite your foster dog to offer a well-loved behavior
to help him recover his emotional equilibrium.

The key to making any of these games work to help your foster be brave is to be sure you keep him far enough away from the scary thing, at first, that his brain is able to click into play mode. You will always be more successful if you start the games when you see low levels of stress, rather than waiting until he's in full meltdown. If he's too stressed or fearful, he won't be able to play. If he'll start to play games with you while the scary thing is at a distance, you'll be able to move closer. If he stops playing and shuts down, you've come too close. Depending on the dog and how fearful he is, you may find some of these play strategies work well enough to walk him past scary stimuli the first time you try, or you may have to work up to it.

*Play with your dog to help him get brave in
the sub-threshold presence of scary stimuli.*

Get Behind. Get Behind is more of a management strategy. Timid dogs often seek to hide when they are afraid. If you teach your foster a cue that means "hide behind me," your body shield can help him get through scary moments.

To teach this behavior:

1. Have the dog in front of you, with an ample supply of small, high-value treats in your treat pouch, or in a bowl on a nearby table.

2. Say "Get Behind!" and lure him behind you and into a Sit. Click and treat.

3. Repeat several times, until he lures easily into position.

4. Now say the cue and pause, to give him a chance to think about it and respond. If he moves even slightly, click, lure him into position, and treat. A tentative movement is sometimes a question to you—"Is this what you want?" If you answer with a hearty "Click (Yes!!)" and treat, you can move the training forward more quickly.

5. Keep repeating the cue/pause, gradually reducing how much you lure, until he's moving into position on his own when you give the cue.

You can start applying this strategy in real-life situations early on in the training, even before your foster dog fully grasps the concept, simply by luring him into his safe position as the scary thing passes. Even with all of the games and management strategies, however, you *must* protect your foster from scary things. If you force him to accept the attentions of bearded men while he is still fearful of them you can *sensitize* him to the scary stimuli, making his behavior worse instead of better.

Treat and Retreat. Treat and Retreat is a procedure to help timid dogs get brave. Its development is attributed to two well-known trainers: Dr. Ian Dunbar, veterinary behaviorist and founder of the Association of Pet Dog Trainers (www.dogstardaily.com/training/retreat-amp-treat), and Suzanne Clothier (www.suzanneclothier.com). While Dr. Dunbar is often attributed with introducing the concept, Clothier is generally credited with popularizing the procedure under the name "Treat and Retreat."

To use Treat and Retreat, start with your foster dog a safe distance from a person who worries him. Have that person toss a piece of *low-value* kibble over your foster's head. The dog will turn and walk away to get the kibble, then turn back to look at the scary person. When he turns back, have the person toss a *high-value* treat in front of the dog, in the approximate place the dog was originally. (You may want to use some kind of visible target to help your tossing person's aim.)

When the dog comes forward and eats the high-value treat, have the person toss another low-value treat *behind* the dog, then another high-value treat in the original spot. As your foster dog gets more relaxed about coming forward for the high-value treat, have the tosser gradually decrease the distance, so the dog is moving closer to the scary person to eat the treat. If you see increased signs of reluctance with the decreased distance, you've decreased the distance too quickly. Go as slowly as necessary to keep the dog happy about this game—you want him moving toward the person tossing the treats happily and voluntarily.

Alphabet soup: CAT/BAT/LAT

In addition to counter conditioning, there are other well-developed protocols available to help timid dogs gain confidence. Constructional Aggression Treatment (CAT) developed by Kellie Snider and Dr. Jesus Rosales Ruiz at the University of North Texas, uses operant conditioning (negative reinforcement) and shaping (dog does deliberate behavior to influence his environment) to convince a dog that his old behavior, in this case acting fearful, no longer works to make a scary thing go away. In the presence of a scary stimulus, the smallest sign of relaxation or confidence now makes the scary thing go away—until the dog learns that acting confident (and *becoming* confident as a result) is a better behavior strategy. Watching their DVD, *Constructional Aggression Treatment; CAT for Dogs*, can help you understand this procedure if you are considering using it.

Behavior Adjustment Training, or BAT, is similar to CAT in some ways, but focuses on having the dog move away from the scary stimulus rather than having the scary thing move away from the dog. Developed by Grisha Stewart, CPDT-KA, CPT, it uses desensitization together with a **functional reward** for calm behavior. You begin at a distance where your dog can see the fear-causing stimulus (scary man with beard) without reacting to it. When your dog offers any form of calm body language you move away from the bearded man as the functional reward. Stewart's book, *Behavior Adjustment Training*, can give you a fuller understanding of the procedure.

BAT defines a "functional reward" as "what your dog wants to happen in that moment." In the case of a fearful dog, what the dog wants is for the scary thing to be farther away. According to Stewart, a good functional reward for a dog's calm behavior can be to move away from the bearded man (negative reinforcement). Similar to CAT, if you teach your dog that calm behavior makes scary things get farther away, your dog will learn to be calm, confident and not fearful in the presence of those things.

LAT stands for "Look At That"—a protocol developed by Leslie McDevitt, CPDT-KA, CDBC, author of *Control Unleashed*, at her training center outside of Philadelphia. In LAT, the key is to keep the dog below threshold (i.e., quiet and calm) while teaching him to look at a scary stimulus, then rewarding him for looking at it. To train LAT, click and reward your foster dog the second he looks at the bearded man, *as long your dog doesn't react adversely*. If the dog is too close to threshold with a bearded man (or other scary stimulus) at *any* distance, start with a neutral target like a yogurt lid or other item the dog has no association with and click as soon as he looks at it. When your foster is offering a quick glance towards the target, name it "Look!" The dog will quickly start to look at his scary triggers when you give the "Look!" cue, and turn back to you for a reward. If your foster dog does not turn quickly, he's probably too close to threshold, or over threshold entirely. Increase the distance between you and the bearded man and try again. Gradually decrease distance as the dog learns to do the "Look!" game with things that are worrisome to him. Look—he's getting braver!

Many of the above games and strategies are compatible with each other. If you are interested in applying the more complex ones, such as CAT, BAT and LAT, you'll need to learn more about them, and enlist the aid of a qualified positive behavior professional. CAT and BAT tend to be mutually exclusive because one moves the dog away from the scary thing, while the other moves the scary thing away from the dog. Other than that, the more of the above strategies you apply, the more tools you'll have at your disposal to help your dog cope with fear-causing stimuli in his world, and the more confident he'll become. Time to get brave!

Coddling and cuddling

There is an unfortunate myth floating around in some parts of the dog training world that if you give reassurance to a fearful dog you will reinforce his fearful behavior. Therefore you must ignore your foster dog when he's trembling at your feet in fear.

Hogwash. Think back in your own life to a time when you were very frightened or upset. Did it help (or would it have helped) you feel better to have someone you trusted come and put his or her arm around your and calmly reassure you that everything was going to be alright? Of course it did. Our dogs are no different.

At times when emotions run high, we are more concerned with helping our dogs get those emotions back under control. In fact, when a dog is very afraid, the emotional part of the brain—the amygdala—takes over, and the thinking part of the brain—the cortex—doesn't work well. The over-threshold dog isn't even capable of connecting his behavior to reinforcement—which is why we try hard in our behavior modification protocols to keep the dog below threshold—so learning can happen. If he is shaking in fear, your calm voice and slow, massage-pressure touch can be hugely reassuring to him. (Fast rubbing and an anxious tone of voice, however, are not.) We even have phrases in English that describe this emotional override: "I was out of my mind with worry." "I was so frightened I couldn't think straight." So the next time your foster dog is emotionally overwrought—trembling at your feet because a thunderstorm is approaching—as long as he finds your calm reassurances truly comforting, you have my blessing.

You have my permission to cuddle your frightened foster. Photo: Shanon McAuliffe

8

Problem Behavior: Aggression and Your Foster Dog

Aggression. It's a natural, normal dog behavior. It's also a scary word that evokes images of maulings, and worse, dog-related fatalities. The term actually covers a wide continuum of behaviors, some of them very appropriate and critically important to successful canine communication. This broad spectrum of behaviors is technically called **agonistic behavior,** defined in ethology as: *"pertaining to the range of activities associated with aggressive encounters between members of the same species or social group, including threat, attack, appeasement, or retreat."* So, while a growl-lunge-bite sequence would be easily recognized by most people as part of the aggression continuum, more subtle agonistic behaviors—such as a freeze, a hard stare or even a lack of eye contact—may not.

Aggression is probably the most common behavioral problem in dogs seen by behavior professionals and/or surrendered to shelters or rescue groups, and the most dangerous one seen in companion dogs. While the number of dog-related fatalities (approximately 30 per year in the U.S.) pales in comparison to accidental death by other means, the number of annual reported bites is staggering. According to the "Dog Bite Law" website (www.dogbitelaw.com): *"The most recent official survey, conducted more than a decade ago, determined there were 4.7 million dog bite victims annually in the USA. A more recent study showed that 1,000 Americans per day are treated in emergency rooms as a result of dog bites. Dog bite losses exceed $1 billion per year, with over $300 million paid by homeowners insurance."*

*There is a continuum of agonistic behaviors that are natural and
normal canine communication signals. You don't have to panic if your
foster dog growls or gives a hard stare—he's trying to communicate!*

Our culture and aggression

Our culture has become oversensitized to dog bites. Once upon a
time if Johnny was bitten by a neighbor's dog, Mom asked Johnny
what he did to the dog. Today she reaches for the phone to call her
attorney. Once upon a time if the family dog snapped at the baby,
Mom learned to be more careful about not letting the baby pull the
dog's ears. Today she's dialing up a behavior professional, or worse,
dropping the hapless dog off at her local shelter. Sensationalized
"dog mauls baby" headlines have turned us into a nation of aggress-
a-phobes, where the smallest indication of discomfort on a dog's part
sends humans screaming for their lawyers.

Behavior professionals mull over the causes of what looks to be a
huge and growing problem. The population shift away from rural
living and toward urban and suburban homes may have lessened
society's understanding of animal behavior in general. This lack of
understanding manifests as inappropriate human behavior toward
dogs, which triggers more aggressive behavior as well as a lower tolerance for bites—even minor ones. A more responsible dog-owning

population is keeping dogs at home rather than letting them wander (which is a good thing!), but as a result dogs may be less well socialized—and more likely to bite. There has been an increase in popularity of dog breeds that contribute to our cultural sensitization—large, powerful breeds who can do serious damage if they bite, such as Pit Bulls and Rottweilers (no, I'm not saying Pits and Rotties are bad, just that when they get into trouble it tends to be big trouble), as well as space-sensitive breeds who have a lower tolerance for inappropriate human behavior, such as Border Collies and Australian Shepherds. The influx of poorly bred puppy mill puppies from breeders who pay no attention to temperament is also a likely factor, as is the trend to rescue and rehome a multitude of unsocialized dogs from hoarder and neglect cases. Finally, the appropriately diligent efforts of animal control authorities to quarantine dogs who bite (for rabies control purposes) and craft "dangerous dog laws" (for public safety purposes) have probably fueled the alarmist reactions to even minor dog bites.

I'm not saying aggression isn't a serious behavior. But there's aggression, and then there's **serious aggression** and it is possible you will encounter it in your foster dog. In a perfect world, all humans would recognize and take appropriate action at the lower levels of agonistic behavior. If that happened, we would rarely see serious aggression—in fact we'd rarely see any bites at all. Until that time, we can only work, one dog and one human at a time, to expand human understanding of canine aggression.

Stress

Across the board, with one tiny exception so rare it's barely worth mentioning, aggression is caused by stress. Whatever "classification" of aggression an owner or behavior professional chooses to use, the underlying cause of the aggression is stress. There is usually a *triggering* stressor—when a dog bites a child, it's a good bet that child was a stressor for him—but there is also a background noise of other stressors that pushed the dog over his bite threshold with *that* child on *that* particular day. These are often stressors that we don't even notice, and because cortisol, an important stress hormone that plays a role in aggression, can stay in the system for at least two to three days, there can be stressors influencing behavior today that occurred yesterday, or even the day before!

Think of it as canine road rage. In the human world, road rage might look like this:

- **Stressor 1**: Our subject jumps out of bed in the morning realizing that his alarm didn't go off and he's late for work.

- **Stressor 2**: He dashes through a cold shower because his hot water's on the blink.

- **Stressor 3**: As he hurries out the door his eye falls on the foreclosure notice that arrived in yesterday's mail because his mortgage payment is overdue.

- **Stressor 4**: He jumps in his car, starts the engine and sees that his gas gauge is on "E." He's already late and now he has to stop to fill up his tank.

- **Stressor 5**: As he pulls onto the freeway his cell phone chirps to remind him of an important meeting in 15 minutes—and his commute is 25 minutes.

- **Stressor 6**: He remembers that his boss warned him that if he's late for one more important meeting he'll be fired. If he speeds, maybe he can get there in time.

- **Stressor 7**: Traffic is a little slow, but if he uses the commuter lane, maybe he can make it. Just as he starts to pull into the left lane a car cuts him off and then pokes along in front of him below the speed limit. It's the last straw. Over threshold, he reaches under his seat pulls out his loaded .357 Magnum and…

In the dog world, canine road rage might look like this:

- **Stressor 1**: Dog has a little isolation distress, usually mitigated by the presence of his canine sibling, but today his brother got dropped off at the vet hospital when his humans went off to work, so he's all alone.

- **Stressor 2**: UPS delivery arrives, and dog has a "thing" about delivery people.

- **Stressor 3**: Just before noon a thunderstorm passes through. Our dog is thunder sensitive, and his owner didn't give him his thunder medication this morning.

- **Stressor 4:** Pet walker is supposed to arrive at 1:00pm, but is late and doesn't get there until 2:30. Dog is stressed by the change in routine and by the urgency of a very full bladder by the time the walker arrives.

- **Stressor 5**: Humans arrive home at their normal time but they are stressed because there are dinner guests due at 7:00pm and they have to get ready. Dog is stressed by his humans' stress, and the fact that they rush though his evening routine, feeding him hurriedly and skipping his walk to the dog park for exercise.

- **Stressor 6**: Visitors arrive, and while the dog is fine with adult visitors, he is not especially fond of children, and there are four in this family. All throughout dinner, the dog listens the high-pitched sounds of children's voices laughing and arguing, and occasionally sees them staring at him.

- **Stressor 7**: After dinner the kids are running around the house. The dog tries to stay out of their way, but eventually one corners him in the kitchen. Over threshold, he pulls out his loaded mouth and…

Stress is an emotional and physiological response to a stimulus. The foundational underpinning of aggression is based on classical conditioning—a dog's emotional and physical response to a stimulus that causes him stress: fear, pain, anger and/or some other strong emotion. He can't help his emotional response any more than you can when faced with something scary or painful. Aggression also has an operant piece—the dog learns that he can deliberately act to make scary stressors go away. When he growls, barks and lunges, perceived bad things tend to leave—so his aggressive behavior is negatively reinforced (dog's behavior makes a bad thing go away), and increases over time.

Aggression is *not* dominance

There's a widespread misconception held by many dog owners, perpetuated by unfortunate television drama, that aggression is all about dominance, and that the appropriate response to any display of aggression is to force the dog into submission. This couldn't be further from the truth. In fact, a very mild, easily resolvable display of aggressive behavior can quickly become a significant behavior problem if the dog's human responds with aggression.

As we discussed in Chapter 5, the concept of dominance in a social group has been so widely misunderstood and distorted that many knowledgeable behavior professionals hesitate to even use the term. In fact, dominance has little to do with aggression, and a lot to do with access to desired resources; the concept of dominance strictly refers to an interaction or a series of interactions between two individuals in which there is an outcome in favor of one member of the pair. That outcome is largely determined by a submissive or yielding response from one of the individuals, *not* through overt conflict or escalated aggression. Someone who is truly higher ranking in social status doesn't *need* to resort to aggression to get what he wants. This holds true for social groups of all species, including humans. Violent behavior between group members is inappropriate and unacceptable in social interactions.

Using violent behavior against a dog who is aggressive adds additional stress to his stress load. You may be able to suppress his aggressive behavior in that moment, but you have likely increased the probability for future, possibly more intense aggression to occur. There are far more appropriate and effective ways to manage and modify aggressive behavior than aggressing back.

So what *do* you do when your foster dog exhibits aggressive behavior? Remember that stress, not any desire to take over the world, causes aggression to erupt. The first thing to do is educate yourself about dog body language so you can be aware of more subtle agonistic behaviors in canines. Then be aware of your foster dog's stressors and stress levels, and avoid putting him in situations where he may be compelled to bite. When you do see stress signals, even subtle ones, remove him from the immediate proximity of the stressor to help him cope with the situation.

When you've identified something that appears to be a stressor for him, figure out how to remove it as a stressor in his life, after noting it in his records so you can pass the information on to your foster organization and potential new owners. If it's something you can get rid of, simply get rid of it. If you can manage it by removing him from the environment when you know the stressor will be present, do it. If it's too present in his world to get rid of or manage, take steps to change his opinion of that stressor through counter conditioning, or change his behavior in the presence of that stressor through operant conditioning. Finally, there will be some low level stressors that he'll just have to live with. As long as they aren't significant enough to put him near or over his bite threshold, he can live with some stressors. We *all* have *some* stress in our lives...

Strategies for eliminating stressors

1. **Get rid of it.** Anything aversive that causes unnecessary pain or stress, including shock collars, choke chains and prong collars, penny cans or throw chains. Even head halters, considered by many to be positive training tools, are aversive to some dogs. Also identify and treat any medical issues that cause pain or discomfort, such as ear infections or arthritis—these are huge stressors.

2. **Manage it.** If your foster dog isn't fond of small children, and if there are *none* in your life and he doesn't encounter them regularly in your neighborhood, you can manage his environment the one time each year your sister comes to visit with your young niece and nephew, by keeping him in another part of the house when the kids are awake and about—and make sure he doesn't get adopted to a home where small children will be present.

3. **Change his association.** Convince him that something that stresses him is actually very wonderful by pairing it consistently with something else wonderful. If your foster dog is stressed by men with beards, you can convince him that men with beards always make chicken happen, by having a bearded man appear, and feeding the dog bits of chicken, over and over and over again, until he *wants* furry-faced men to appear so he can have more chicken. The key to successful counter conditioning is to always keep the dog below threshold; you want him a little aware of and worried about the aversive stimulus, but not quaking in fear or barking and lunging.

4. **Teach him a new behavior.** Perhaps your foster becomes highly aroused by visitors coming to the door. He's not fearful or aggressive, but the high arousal is a stressor. You can teach him that the doorbell is his cue to run and get in his crate, where he'll receive a stuffed Kong or other doggie delectable. Or you can teach him that visitors toss toys for him to chase if he sits politely when the door opens.

> **5. Live with it.** So you're a little (or a lot!) stressed because your work isn't going well, or the school just notified you that your teenage daughter has been skipping school. While I encourage you for your own well-being to take steps to reduce your own stress as much as possible, this is one your dog can live with, especially if you remember that when you are stressed, it pushes your dog a little closer to his own bite threshold.

When aggression happens

What if you misjudge a situation and something happens that puts your foster dog over threshold and causes him to display seriously aggressive behavior, perhaps even biting someone? First, don't panic. All dogs can bite, and the fact that your foster has doesn't make him a Cujo. You will need to:

- Move him away from the scene. Calmly stash him in another room, stick him in your car for a moment, or hand his leash to someone he knows who is not at risk for being bitten and have them take him away. *Do not punish him!*

- Apologize. A good apology is, "Oh, I am so sorry you (or your dog, or your child) were bitten!"

- Examine the site of the bite. Take a couple of quick photos if you can. If the bite broke skin, offer first aid (if you have it). If the injuries are serious, call 911, or see that the victim has a way to access medical care.

- Contact your foster organization as soon as possible. Be prepared to hear that this means they will not be able to place him in a home. Many shelters and rescue groups simply cannot risk the liability and possible financial destruction of their organization if they rehome a dog with a known bite history, the dog subsequently bites again, and they are sued.

- Follow your organization's policies regarding contact and communication with the bite victim. A preemptive discussion with your group about this possibility is a good idea, to prevent yourself from taking on more liability than is appropriate, while still doing the ethical and reasonable thing. Be wary of immediately accepting responsibility for the incident; talk with your organization and/or your attorney first.

- Prepare for a visit from animal control. These days, in most parts of the country, if a dog bite breaks human skin the dog must be quarantined for at least ten days. Have your current rabies certificate handy. They *will* ask to see it; a rabies tag isn't enough. In many jurisdictions you may be able to quarantine the dog in your own home. If not, find out if the dog can be kept at a vet hospital for the required period—it's usually a safer, less stressful place than a shelter. If animal control insists on taking the dog away for quarantine, *do not sign anything* until you have checked in with your organization, read the document carefully and are sure you understand it. Some caretakers have unknowingly and tragically signed their dogs over for euthanasia when they thought they were just agreeing to quarantine. *Note: Your group may not want you to sign anything—they may want to deal with animal control themselves.*

- Prepare for "dangerous dog proceedings." Depending on the laws in your area, your dog may be declared "potentially dangerous" for acting aggressive, or "dangerous" for actually biting someone. It's good to read your local ordinance now, even if your dog never bites anyone, but for sure *after* a bite happens. If your foster dog is designated "dangerous" or you do get called to a hearing of some kind in relation to the dog's aggressive behavior, you'd be wise to involve an attorney.

You would like to never have to post this sign on your fence.

Preventing aggression

Basic training and early socialization can go a long way toward inoculating your foster dog against future aggression. Your observational skills and ability to mitigate stressful situations for him are excellent booster shots. At that point, however, when you become aware that your foster dog's behaviors are inappropriate, travelling along that continuum of agonistic behavior verging on overt aggression, and are resistant to your efforts to manage and modify them, it's time to call for help.

Remember that a good behavior professional won't come riding in like a white knight, push your foster dog around a little and declare him cured. A good behavior modification protocol is not dramatic, but rather a slow, low-key program that will help your foster learn to cope with his world. Your behavior professional won't need to see the actual aggressive behavior—she will trust your description of the dog's reaction to the stressors in his world and her own observations of his stress signals, and help you figure out how to keep him far below his bite threshold. Like most behaviors, aggression is far easier to modify sooner, before the dog has had time to practice and get good at it. Finally, regardless of how successful you are at modifying your foster dog's behavior, full disclosure to potential adopters is a vital and ethical imperative, to ensure his future well-being and safety, as well as that of the humans around him.

A good behavior modification program for foster dog aggression involves slow, steady progress. There are no instant fixes or magic wands.

9
Problem Behavior: Separation and Isolation— A Foster Dog's Worst Nightmare

The term "separation anxiety" is a pretty mild label for a devastating and destructive behavior that can result in human frustration, anger, and sometimes even the euthanasia of an offending dog when a despairing caretaker reaches her wits' end. If you've ever had the misfortune of walking into your house to find overturned furniture, inches-deep claw gouges on door frames, blood-stained tooth marks on window sills, and countless messages on your answering machine from neighbors complaining about your dog barking and howling for hours on end in your absence, you're probably familiar with the term. Thirty years ago the phrase was uncommon in dog training circles. Today it's a rare dog owner who hasn't heard of separation anxiety, experienced it with a personal or foster dog, or at least had a dog-owning friend whose canine companion reportedly suffered from this difficult disorder.

Veterinary behaviorist Dr. Karen Overall defines **separation anxiety** as "A condition in which animals exhibit symptoms of anxiety or excessive distress when they are left alone." Common signs of the condition in the absence of the owner include:

- Destructive behavior

- House soiling

- Excessive vocalization

Many dogs with this challenging behavior also:

- Refuse to eat or drink when left alone

- Don't tolerate crating

- Pant and salivate excessively when distressed
- Go to great lengths to try to escape from confinement, with apparent total disregard for injury to themselves or damage to their surroundings

It's natural for young mammals to experience anxiety when separated from their mothers and siblings; it's an adaptive survival mechanism. A pup who gets separated from his family cries in distress, enabling Mom to easily find him and rescue him. In the wild, even an adult canine who is left alone is more likely to die, either from starvation, since he has no pack to hunt with, or from attack, since he has no packmates for mutual protection. Given the vital importance of a dog's canine companions, it speaks volumes about their adaptability as a species that we can condition them to accept being left alone at all! We're lucky we don't have far more problems than we do, especially in today's world, where few households have someone at home regularly during the day to keep the dog company.

It's natural for dogs to experience some level of distress when separated from their social group. We need to help them understand that it's okay to be alone.

There was a time in our society when fewer dogs were left home alone—Mom stayed home while Dad went off to work every day—so dogs had less exposure to the kind of daily isolation that contributes to separation anxiety behavior. Some behavior scientists theorize that experiencing a fear-causing event when a young dog is already mildly stressed about being alone can trigger more intense "home alone" anxiety behaviors. In today's world there are a significant number of dogs who are afflicted with some degree of separation distress. Sadly, many of these dogs end up in shelters and with rescue groups, and it means that many foster dogs suffer from it.

Differential diagnosis

Another reason separation anxiety *seems* so prevalent these days compared to a few decades ago is that it is misdiagnosed with some frequency by laypersons. With an increased awareness of the condition has come an increase in misidentification of behaviors that resemble separation distress behaviors, but really aren't.

For example, house soiling *can* be related to anxiety, but the cause could also be incomplete housetraining, lack of access to appropriate elimination areas with unreasonable owner expectations (expecting the dog to "hold it" for ten hours or more), fear, excitement, marking, submissive elimination or physical incontinence. Destructive behavior may be a result of separation anxiety, or it could be normal puppy behavior, play, reaction to outside stimuli, and/or an outlet for excess energy. Separation distress could be the cause of excessive barking and howling, or the dog could be stimulated to bark by street sounds (traffic, people talking), trespassers (i.e., mail carrier, intruder, Girls Scouts selling cookies), social facilitation (other dogs barking), play, aggression or fear.

It's critically important that a problem behavior be correctly identified prior to the implementation of a behavior modification program. It does no good to try to modify separation anxiety if that's not really the problem.

If elimination accidents occur when the owner is home as well as when the dog is left alone, it's more likely a housetraining problem than a separation issue. Separation-related destruction is usually directed toward escape efforts—chewing or clawing at or through doorframes, window sills and walls. If the destruction is more generalized throughout the house, it points toward one or more of the other possible causes, rather than an isolation issue. A strategically located video camera or sound-activated tape recorder can help identify possible outside stimuli, such as visitors to the home or unusual noises, that might trigger what otherwise may appear to be separation-related behaviors.

A continuum

Distress over being left alone is not always a full-blown separation anxiety problem. A dog may suffer from mild to severe *isolation* distress or anxiety or mild to severe *separation* distress or anxiety.

The difference between distress and anxiety is a matter of degree on a continuum. "Distress" indicates a lower intensity of stress behaviors when the dog is alone, while "anxiety" is an extreme panic attack. This distinction between "isolation" and "separation" is equally important. *Isolation* distress means the dog doesn't want to be left alone—any ol' human will do for company, and sometimes even another dog, or some other species, such as a cat—will fill the bill. True *separation* distress or anxiety means the dog is hyper-bonded to one specific person, and continues to show stress behaviors if *that person* is absent, even if other humans are present.

Behavior modification

There are a number of steps you can take to resolve your foster dog's isolation or separation anxiety behavior. The program spelled out in the accompanying sidebar, "Preventing Separation Anxiety," can also be used to modify an existing isolation/separation condition. However, you will progress much more slowly through the steps of the program with a dog who already suffers from separation-related behaviors; your foster dog's existing strong emotional response to being left alone will make this a much more challenging proposition.

Here are some other avenues to explore, to complement your modification work:

- Exercise your foster well before you leave. A tired dog has less energy with which to be anxious and destructive. Be sure to end your exercise session 20 to 30 minutes before you go, so he has time to settle down.

- Five minutes before you leave, give him a well-stuffed Kong to take his mind off your imminent departure.

- Make your departures and returns completely calm and emotionless. No huggy/kissy "Mummy loves you" scenes. If he gets excited and jumps all over you when you return, ignore him. Turn your back and walk away. When he finally settles down, say hello and greet him very calmly.

- Defuse the pieces of your departure routine by also doing them when you are not leaving. Pick up your car keys and sit down on the sofa to watch TV. Dress in your business suit and then cook dinner. Set your alarm for 5:00am on a Saturday, then roll over and go back to sleep.

- Mix up the pieces of your departure routine when you *are* leaving, so his anxiety doesn't build to a fever pitch as he recognizes your departure cues. We are creatures of habit too, so this is hard to do, but can pay off in big dividends. Eat breakfast before you shower instead of after. Pick up your keys and put them in your pocket before you take your foster out for his final potty break. Put your briefcase in the car while you are still in your bathrobe. Make the morning ritual as unpredictable as possible.

- Use a "safe" cue such as "I'll be back," *only* when you know you'll return within the time period your dog can tolerate. As suggested in Patricia McConnell's wonderful booklet on separation anxiety titled *I'll Be Home Soon*, this helps your foster dog relax, knowing he can trust you to return.

- Explore alternative dog-keeping situations to minimize the occasions when you do have to leave him alone—doggie daycare may be suitable for some dogs, but not for others. You may be able to find a neighbor or relative who is house-bound and might appreciate some canine companionship. If your workplace is dog-friendly, take him to work with you!

- Use Comfort Zone/Adaptil (DAP) plug-ins and sprays in his environment to help ease his anxiety.

- Play the *Through a Dog's Ear* CDs—soothing music that helps calm a stressed dog (or human!). (Play it frequently when you *are* home as well as when you leave, so it doesn't become a predictor to the dog that he's about to be left alone.)

- Remove as many other stressors from your foster dog's world as possible to help him maintain his equilibrium in your absence. No choke chains, shock collars, physical or harsh verbal punishment (especially in connection to his anxiety behaviors).

- Consider working with a behavior professional to be sure you're on the right path. A good behavior professional can also help you explore the possibilities of using anti-anxiety medications to maximize the effectiveness of your modification efforts.

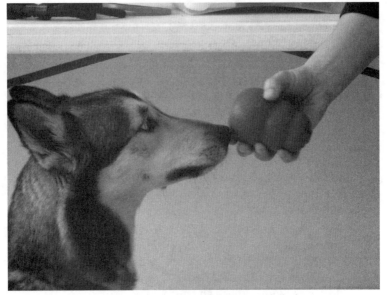

Give your dog a well-stuffed Kong to help him
decide that it's good to be left home alone.

Fixing true separation anxiety is hard work. It's all too easy to get frustrated with your foster dog's destructive behavior. Remember that this behavior may well be why he's homeless in the first place. Also remember that he's not choosing to do it out of spite or malice—he is panicked about his own survival without you, his pack, there to protect him. It's not fun for him either—he lives in the moment, and the moments that you are gone are long and terrifying. If you make the commitment to modify his behavior and succeed in helping him be brave about being alone, you'll not only save your home from destruction, you will enhance the quality of your foster dog's life immensely—as well as your own—and perhaps save him from destruction too.

Preventing separation/isolation anxiety

The most important ingredient in a successful separation/isolation anxiety prevention program is to set your foster up for success. When you bring a new foster home, implement a program to help him be comfortable with being alone for gradually increasing periods. This will help to assure him that it's not necessary to panic—he hasn't been abandoned yet again; you always come back. Be sure to exercise him well before you practice—a tired dog is a much better candidate for relaxation than one who's "full of it."

Here are the ten steps of a two-day program to create a dog who is comfortable being left alone (one who is not already suffering from separation/isolation anxiety):

1. **Bring your dog home** at a time when someone can spend a few days with him to ease the stress of the transition.

2. **Prepare a quiet, safe space** in advance such as a playpen or puppy pen, or a dog-proofed room such as a laundry room.

3. **When you bring your dog home**, give him a chance to relieve himself outdoors, and spend ten to fifteen minutes with him in the house under close supervision. Then put him in his pen and stay in the room with him.

4. **Stay close at first**. Read a book. If he fusses, ignore him. When he's quiet, greet him calmly, take one step away, and then return before he has a chance to get upset. Speak to him calmly, then go back to reading. You're teaching him that if you leave, you will return. Other family members should make themselves scarce during this time—your dog needs to learn to be *alone*.

5. **Continue to occasionally step away**, gradually increasing the distance and varying the length of time that you stay away, so that eventually you can wander

around the room without upsetting your dog. Each time you return, greet him *calmly*. Every once in a while say "Yes!" in a calm but cheerful voice before you return to him, then walk back to the pen and feed him a treat.

6. **After an hour or so, give him a break**. Take him outside to potty and play. Hang out for a while. Then go back inside and resume his pen exercises.

7. **Begin again, staying near the pen until he settles**. More quickly this time, move along steps four and five until you can wander around the room without generating alarm. Now step into another room very briefly, and return before your dog has time to get upset. Gradually increase the length of time you stay out of the room, interspersing it with wandering around the room, sitting near him reading a book, and sitting across the room reading a book. If he starts to fuss, wait until he stops fussing to move back toward him. Teach him that calm behavior makes you return, fussing keeps you away.

8. **Occasionally, step outside of the house**. Your goal for day one is to get your foster comfortable with you being away from him for fifteen to twenty minutes. (It's usually the first twenty minutes of separation that are most difficult.) Vary the times, so he doesn't start getting antsy in anticipation of your return. Remember to give him plenty of potty and play breaks, every hour for a young pup, every one to two hours for an older dog.

9. **On day two, quickly go through the warm-up steps again**, until you can step outside for fifteen to twenty minutes at a time, interspersed with shorter separations. On one of your outdoor excursions, hop into your car and drive around the block. Return in five to ten minutes, and calmly re-enter the house just as you have been during the rest of the exercises. Hang out for a while, then go outside and drive away again, for a half hour this time.

> **10. Now it's time for Sunday brunch.** Be sure your foster dog gets a thorough potty break and play time, then give him fifteen minutes to relax after the stimulation of play. Put his favorite stuffed Kong into his pen, round up the family, and calmly exit the house for an outing of a couple of hours' duration. When you arrive home to a calm and happy dog, drink an orange juice toast to your graduation from Separation Anxiety Prevention School.
>
> *Note—If you are modifying an already existing distress or anxiety condition you will need to work through the steps of the program much more slowly.*

It's unfair to ask a young dog or any newly-arrived foster to stay home alone for eight to ten hours—he needs to get out to relieve himself midway through the day. If you force him to soil the house or his crate, at worst you can cause stress-related behaviors, at best you may create housetraining problems. Options might include taking him to work with you, having family members come home on their lunch hour, arranging for stay-at-home neighbors to take him out, hiring a pet walker to walk him and play with him, or sending him to a well-run doggie daycare environment. *(Note: the daycare option is not appropriate for a very young pup, or for a dog with true separation anxiety.)* If you set up a routine to help your foster dog succeed, he'll someday excel in his Home Alone studies, and you will have greatly enhanced his prospects for success in his forever home.

10

Placing Your Foster Dog—It's Okay to Cry

The time has come. You've done your job, and done it well. Your foster dog's broken leg has healed, the upper respiratory infection has resolved, the puppies are well socialized and ready for placement, the behavioral issues are well on their way to being resolved. Whatever the reason he was in foster, it's time for him to go to his forever home. Your parting will be bittersweet. Your joy at seeing him go on to his new life will likely be tempered by the sadness of your own loss.

It's hard to say goodbye.

Finding his forever home

Your partner organization may require you to bring your foster back to their shelter for adoption, or you may be able to help with his placement while he remains in your care. Clearly the latter is preferable from *his* point of view, but it may mean he doesn't get the same exposure he might if he were in a kennel, interacting with potential adopters as they walk past. Still, exposure or no, dogs who are very stressed by a shelter/kennel environment would do well to stay in their foster homes while their forever homes are found.

If you have a choice, I'd suggest keeping your foster dog with you and taking advantage of any adoption events you can take him to, as well as print advertising and social media. This is one of those questions you hopefully had answered before you signed on with this group, so there are no last minute surprises.

Most rescues and shelters are well aware of the various options for promoting adoptions. The large chain pet supply stores hold regular adoption events where groups can bring their adoptable animals to meet and greet potential adopters. Many groups post their adoption dogs on Facebook. And there are sites like Pet Finder and Petango where non-profit groups can post photos and information about their adoptable dogs. Talk with your partner organization about other ways, like ads in the classified section of your local newspaper, that you might be able to promote the adoption of your foster.

Best-case scenario

You've found an appropriate placement for your foster dog, or your foster organization has. The potential adopter has been thoroughly screened, including, perhaps, a home visit. The entire family has met your foster dog, and he loves them all—and they love him. This should be the easy part...but letting go can be hard.

Don't be surprised if you feel the loss, and grieve, in the same way you might when you lose an animal companion to death. Your grief will be mitigated somewhat by knowing your well-loved foster charge is moving on to a new life in his forever home—but you will likely still grieve.

Letting go. You and your foster have likely bonded during his time with you. It could be hard for both of you to let go—and yet you must. There are lots of "failed fosters" out there—the ones who are ultimately adopted by their foster humans—and chances are you'll have some of these yourself. Most foster parents do, sooner or later. Still, you can't keep *all* of your fosters, and the more of them you keep, the fewer you'll be able to help in the future.

Depending on the policies of your foster organization, you may be able to stay in touch with your foster's new family. If so, do if you want—but don't be a pest. Let his new humans know you are willing to be a resource for them if they have questions, and ask them to keep you posted on his adjustment, but refrain from bugging them as often as you might want. Computer communication—e-mails or Facebook—might be preferred over telephone calls or in-person visits; ask them how—and how often—they are comfortable having you communicate. Then honor their preferences. Don't be a stalker.

You may want to create a goodbye ritual for yourself—dinner with good friends who understand, a long hike in the hills, or some other activity that helps you celebrate your success, and your foster dog's happy future.

Visiting. Give your foster's new family at least a month, preferably two or more, before you even think about visiting. They need time to bond before you renew your own acquaintance. With luck, by then a new foster has come into your world to ease the pain of loss, and your heart is happily engaged with your new foster project. If the adopters are willing, meet in a public place for a brief, but happy encounter, and then go on about your separate ways. Stay in touch online, watch for the occasional Facebook photo, and know that you did a very good thing. Then focus on doing the same for your next foster.

Worst-case scenario

It happens. Not too often, thankfully, but it happens. You discover some significant health or behavior issue with your foster dog that makes him impossible to rehome.

Health. Perhaps your foster has been diagnosed with a terminal illness while in your care, and he doesn't have much time left. Not fair to ask an adopter to take him on, and perhaps not fair to *him* to uproot him yet again for the brief time he has left, in yet another new home. You might choose to keep him yourself, to give him the best care possible for whatever time he has left. Or you might not—and that's okay.

If he's suffering despite medical care, or there is no one who can provide him with end-of-life care, you and the organization may together decide it's time to let him go, and humanely end his life. You should have the option of being with him if and when that time comes, to make his passing as easy and stress-free as possible. If you can't bear the thought of being with him, if you know that your emotions will make it harder for him, then you can return him to the compassionate care of the shelter or rescue workers who entrusted him to you, and know that they will help him to a gentle death. Letting go, as much as it hurts your heart, however you choose to do it, is always the right answer. The only wrong choice is one that causes him to suffer unnecessarily because you can't bring yourself to let go.

Behavior. You may also discover that your foster dog has significant behavioral issues that, despite your best efforts and intentions, have not responded to behavior modification interventions. Again, you may choose to adopt him yourself—if your organization allows you to do so—knowing you are taking on the potential liability of a dog who may bite someone, or the challenge of a dog who is seriously destructive. If you are confident that you can manage his behavior, continue to work to modify it, and provide him with a good quality of life, then kudos to you—and congratulations to him.

You may also choose to let go—and that's a right answer also. You are not obligated to offer your home to a dog who is damaged beyond repair. You have a life to lead as well, and you aren't required to sacrifice your quality of life for a dog who is facing years of stress and unhappiness himself. It won't be easy to have your foster euthanized for behavioral reasons, and you may always second guess yourself, but it can be a right and appropriate choice. Those who are faced with the difficult decisions that are inherent in working with homeless dogs will understand and support you.

Here are some wise words posted online from a rescue foster parent: "I've had to have rescues put down, I've had to have my own beloved dogs put down. It was never easy for me, but the knowledge that the dog was no longer suffering made it bearable. I cannot look at a suffering dog and feel that it's kinder to watch him suffer."

My foster dog: New York Squiddy

Our little Squid from Chapter One was a rousing foster success. After six weeks of multiple daily hikes on the farm, trainers working with him regularly to help him learn tolerance for frustration and impulse control as well as basic good manners, and two weeks of intensive training at our dog trainer academies, he was ready to find his forever home.

I contacted Jack Russell Terrier Rescue, and as helpful as they were, none of the adopters they referred to me were quite right for this feisty little boy. We waited.

I posted him on Facebook as available for adoption, and Lydia De Rosche, a positive reinforcement trainer in New York City, e-mailed me to say she had the perfect home for Squid—a client who had recently lost her Doberman to old age, and was looking for a smaller dog for apartment living. She assured me that Claudia Husemann would provide an exceptional home for our little foster boy. The trainer had a good reputation, so I suggested Claudia drive down to meet Squid. My husband Paul was dubious. "He's a country dog," Paul protested. "He doesn't want to live in a big city!"

Claudia drove the four-plus hours from New York City, knowing that even if she wanted him, she wouldn't be able to take Squid back home with her right away. She would need to complete the adoption paperwork, go through the screening process, go back home, wait for him to be neutered, and return to pick him up. Undaunted, she came anyway. It was love at first sight for both dog and human. Claudia would ultimately complete the adoption, return for him, and take him home.

Our trust in Claudia was well placed. She adores her little dog, and he adores her. He goes to work with her and happily gazes out the window of her high-rise office. Our Squid would go on to become widely known as "New York Squiddy"—the subject of several magazine and newspaper articles, happily hobnobbing with celebrities in Central Park, and developing quite a following on Facebook. Total strangers walk up to Claudia on the sidewalks of New York and ask, "Is that Squid?"

I cried as I watched Claudia drive away with our little Squid crated in the back of her car. I saw tears in Paul's eyes as well. It was bittersweet to say goodbye, but I knew he was going to a great home and a great life, and that his departure opened the door for another foster at the Miller home. That's what fostering is about—saving lives, one Squid at a time.

New York Squiddy. Photo: Claudia Husemann

Glossary

Agonistic behavior: Pertaining to the range of activities associated with aggressive encounters between members of the same species or social group, including threat, attack, appeasement or retreat.

Anxiety: The distress or uneasiness of mind caused by apprehensive *anticipation* of *future* danger or misfortune, real or imagined.

Behavior chain: A series of behaviors linked together in a continuous sequence by cues, and maintained by a reinforcer at the end of the chain. Each cue serves as the marker and the reinforcer for the previous behavior, and the cue for the next behavior.

Classical conditioning: Behavioral response based on associations between stimuli.

Cognition: The ability to think, to process thought, to understand concepts and have an awareness of the thought processes of *other* beings.

Conspecifics: Others of the same species.

Dominance theory: The erroneous approach to canine social behavior based on a study of captive zoo wolves conducted in the 1930's and 1940's by Swiss animal behaviorist Rudolph Schenk.

Extinction burst: Presentation of a more intense (louder, longer, harder, bigger, more frequent) level of the behavior you are trying to extinguish, due to the dog's frustration at not receiving reinforcement for a previously reinforced behavior.

Fear: A feeling of apprehension associated with the *presence or proximity* of an object, individual, or social situation.

Functional reward: What your dog wants to happen in that moment.

Long-line: A 15- to 50-foot leash that gives your dog more freedom to run, safely and legally.

Neophobia: Fear of new sights, sounds, and experiences.

Operant conditioning: Deliberate behavior response based on expected consequences.

Parting stick/breaking stick: A carved wooden hammer handle, tapered to a rounded point at one end, inserted between a dog's teeth and turned sideways to pry open the jaws of fighting dogs. Most commonly used with breeds of dogs purposely used for fighting.

Phobias: Persistent, extreme, inappropriate fear or anxiety responses, far out of proportion to the level or nature of threat presented.

Predatory drift: When a larger dog perceives a small running dog as a prey object such as a bunny or squirrel instead of a conspecific, and shifts from play to food-acquisition mode, sometimes with tragic results.

Reflexive behavior: Automatic unconditioned behavior in response to a stimulus.

Separation/isolation anxiety: A condition in which animals exhibit symptoms of anxiety or excessive distress when they are left alone.

Socialization: Giving a puppy *positive* exposures to the world while he's young enough (three weeks to fourteen weeks) to be forming his world view, to convince him the world is a safe and happy place.

Targeting: Teaching your dog to touch a designated body part to a designated target.

Umbilical approach: A method for housetraining puppies and adult dogs in which dog is always either in a crate or pen, on a leash held in hand or clipped to belt (or restrained nearby on a tether, or in the same room as the caregiver with the door closed), under the direct supervision of an adult or responsible teen, or outdoors, until the dog can be trusted with house freedom.

Recommended Reading
and Other Resources
Books and DVDs

The Bark Stops Here. Terry Ryan, 2000, Legacy Canine & Behavior.

Basic Good Manners; A Seven-Week Course DVD. Pat Miller, 2009, Peaceable Paws.

Behavior Adjustment Training. Grisha Stewart, 2011, Dogwise Publishing.

Canine Body Language. Brenda Aloff, 2005, distributed by Dogwise Publishing.

The Cautious Canine; 2nd edition. Patricia McConnell, McConnell Publishing Ltd.

Constructional Aggression Treatment; C.A.T. for Dogs, DVD. Kellie Snider and Jesus Rosalez-Ruiz, 2007, Tawzer Dog Videos.

Control Unleashed. Leslie McDevitt, 2007, Clean Run Productions.

The Culture Clash. Jean Donaldson, 2005, Dogwise Publishing.

The Dog Trainer's Complete Guide to a Happy, Well-Behaved Pet. Jolanta Benal, 2011, St. Matin's Griffin.

Dominance in Dogs; Fact or Fiction. Barry Eaton, 2010, Dogwise Publishing.

Dominance Theory and Dogs. James O'Heare, 2008, distributed by Dogwise Publishing.

Don't Leave Me. Step-by-Step Help for your Dog's Separation Anxiety. Nicole Wilde, 2010, Phantom Publishing.

Don't Shoot the Dog, 2nd Edition. Karen Pryor, 1999, Bantam Books.

Do Over Dogs. Pat Miller, 2010, Dogwise Publishing.

The Genius of Dogs. Brian Hare and Vanessa Woods, 2013, Dutton Adult.

Help for Your Fearful Dog; A Step-by-Step Guide to Helping Your Dog Conquer His Fears. Nicole Wilde, 2006, Phantom Publishing.

I'll Be Home Soon; How to Prevent and Treat Separation Anxiety. Patricia McConnell, 2009, McConnell Publishing Ltd.

The Language of Dogs DVD. Sarah Kalnajs, 2006, Blue Dog Training and Behavior, distributed by Dogwise Publishing.

Manual of Clinical Behavioral Medicine for Dogs and Cats. Karen Overall, 2013, Elsevier.

Oh, Behave! Jean Donaldson, 2008, Dogwise Publishing.

The Other End of the Leash. Patricia McConnell, 2002, Ballantine Books.

Peaceable Paws Good Manners Class Book. Pat Miller, 2008, Peaceable Paws.

Play With Your Dog. Pat Miller, 2008, Dogwise Publishing.

Positive Perspectives. Pat Miller, 2004, Dogwise Publishing.

Positive Perspectives 2. Pat Miller, 2008, Dogwise Publishing.

The Power of Positive Dog Training, 2nd edition. Pat Miller, 2008, Howell Book House.

Real Solutions to Canine Behavior Problems DVD. Pat Miller, 2012, Tawzer Dog Videos.

Through a Dog's Ear CDs, (several). Joshua Leeds and Lisa Spector.

The Whole Dog Journal, monthly magazine, Belvoir.

About the Author

Pat Miller is a Certified Behavior Consultant, Canine (KA), Certified Dog Behavior Consultant, Certified Professional Dog Trainer (KA) and past president of the Association of Pet Dog Trainers (US). Miller offers group good manners classes, private training and behavior modification services, dog training workshops and trainer academies at her Peaceable Paws 80-acre training facility in Fairplay, Maryland, where she and her husband Paul live with their four dogs, three-and-a-half cats, six horses, four chickens, and a pot-bellied pig. In addition, Miller presents seminars and workshops around the world on a variety of training and behavior topics. She has authored five books on dog behavior and training (this is her sixth): *The Power of Positive Dog Training, Positive Perspectives, Positive Perspectives 2, Play With Your Dog,* and *Do-Over Dogs.* Miller is training editor for *The Whole Dog Journal,* also writes for several other publications, and is currently on the Board of Directors for the Certification Council for Professional Dog Trainers. www.peaceablepaws.com

Index

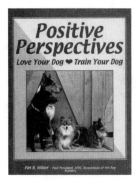

Play With Your Dog
Subtitle
Pat Miller

For most dogs, play comes naturally, while for others, play is something they need to be taught. Play helps dogs learn to interact properly with other dogs—and people. It helps sharpen their social skills and provides excellent physical and mental stimulation. In addition to being just plain FUN, play is a great way to help build a solid relationship between you and your dog and can be a great training tool. Pat includes dozens of game ideas collected from trainers all over the country you can try out with your dog(s).

Do Over Dogs
Give Your Dog a Second Change at a First Class Life
Pat Miller

What exactly is a Do-Over Dog? It might be a shelter dog you're working with to help her become more adoptable. Perhaps it's the dog you've adopted, rescued, or even found running stray who is now yours to live with and love...forever. Or it could be the dog you've lived with for years but you realize he still has "issues" that make him a challenging canine companion. A Do-Over Dog is any dog that you think needs—make that deserves—a second chance in life.

Dogwise.com is your source for quality books, ebooks, DVDs, training tools and treats.

We've been selling to the dog fancier for more than 25 years and we carefully screen our products for quality information, safety, durability and FUN! You'll find something for every level of dog enthusiast on our website www.dogwise.com or drop by our store in Wenatchee, Washington.